Duxbury

Student Solutions Manual

for

Mendenhall, Beaver, and Beaver's
Introduction to Probability and Statistics

Twelfth Edition

Barbara M. Beaver
University of California, Riverside

THOMSON

BROOKS/COLE

Australia • Canada • Mexico • Singapore • Spain • United Kingdom • United States

Printed in Canada
2 3 4 5 6 7 09 08 07

Printer: Webcom

ISBN-13: 978-0-534-46325-0
ISBN-10: 0-534-46325-8

For more information about our products, contact us at:
Thomson Learning Academic Resource Center
1-800-423-0563

For permission to use material from this text or product, submit a request online at
http://www.thomsonrights.com.
Any additional questions about permissions can be submitted by email to **thomsonrights@thomson.com.**

Thomson Higher Education
10 Davis Drive
Belmont, CA 94002-3098
USA

Asia (including India)
Thomson Learning
5 Shenton Way
#01-01 UIC Building
Singapore 068808

Australia/New Zealand
Thomson Learning Australia
102 Dodds Street
Southbank, Victoria 3006
Australia

Canada
Thomson Nelson
1120 Birchmount Road
Toronto, Ontario M1K 5G4
Canada

UK/Europe/Middle East/Africa
Thomson Learning
High Holborn House
50–51 Bedford Road
London WC1R 4LR
United Kingdom

Latin America
Thomson Learning
Seneca, 53
Colonia Polanco
11560 Mexico
D.F. Mexico

Spain (including Portugal)
Thomson Paraninfo
Calle Magallanes, 25
28015 Madrid, Spain

Table of Contents

1: Describing Data with Graphs

1.1   **a**   The experimental unit, the individual or object on which a variable is measured, is the student.
    b   The experimental unit on which the number of errors is measured is the exam.
    c   The experimental unit is the patient.
    d   The experimental unit is the azalea plant.
    e   The experimental unit is the car.

1.7   The population of interest consists of voter opinions (for or against the candidate) <u>at the time of the election</u> for all persons voting in the election. Note that when a sample is taken (at some time prior or the election), we are not actually sampling from the population of interest. As time passes, voter opinions change. Hence, the population of voter opinions changes with time, and the sample may not be representative of the population of interest.

1.9   **a**   The variable "reading score" is a quantitative variable, which is probably integer-valued and hence discrete.
    b   The individual on which the variable is measured is the student.
    c   The population is hypothetical – it does not exist in fact – but consists of the reading scores for all students who could possibly be taught by this method.

1.13   **a**   The percentages given in the exercise only add to 94%. We should add another category called "Other", which will account for the other 6% of the responses.
    b   Either type of chart is appropriate. Since the data is already presented as percentages of the whole group, we choose to use a pie chart, shown in the figure below.

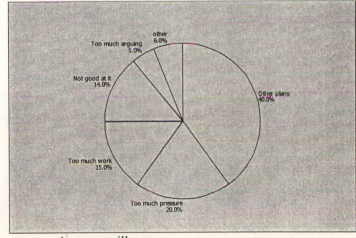

    c   Answers will vary.

1.15   **a**   The total percentage of responses given in the table is only $(40+34+19)\% = 93\%$. Hence there are 7% of the opinions not recorded, which should go into a category called "Other" or "More than a few days".

      b   Yes. The bars are very close to the correct proportions.

      c   Similar to previous exercises. The pie chart is shown below. The bar chart is probably more interesting to look at.

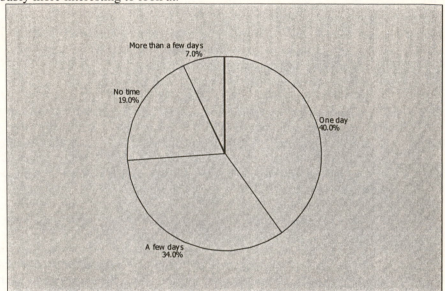

1.21   **a**   Since the variable of interest can only take the values 0, 1, or 2, the classes can be chosen as the integer values 0, 1, and 2. The table below shows the classes, their corresponding frequencies and their relative frequencies. The relative frequency histogram is shown below.

Value	Frequency	Relative Frequency
0	5	.25
1	9	.45
2	6	.30

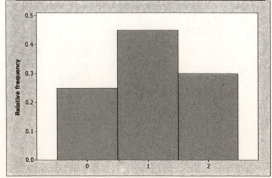

      b   Using the table in part **a**, the proportion of measurements greater then 1 is the same as the proportion of "2"s, or 0.30.

c The proportion of measurements less than 2 is the same as the proportion of "0"s and "1"s, or $0.25 + 0.45 = .70$.

d The probability of selecting a "2" in a random selection from these twenty measurements is $6/20 = 30$.

e There are no outliers in this relatively symmetric, mound-shaped distribution.

1.25   **a**   The test scores are graphed using a stem and leaf plot generated by *Minitab*.

Stem-and-Leaf Display: Scores
```
Stem-and-leaf of Scores  N  = 20
Leaf Unit = 1.0

  2   5   57
  5   6   123
  8   6   578
  9   7   2
 (2)  7   56
  9   8   24
  7   8   6679
  3   9   134
```
b-c   The distribution is not mound-shaped, but is rather bimodal with two peaks centered around the scores 65 and 85. This might indicate that the students are divided into two groups – those who understand the material and do well on exams, and those who do not have a thorough command of the material.

1.31   **a**   The data ranges from .2 to 5.2, or 5.0 units. Since the number of class intervals should be between five and twenty, we choose to use eleven class intervals, with each class interval having length 0.50 ($5.0/11 = .45$, which, rounded to the nearest convenient fraction, is .50). We must now select interval boundaries such that no measurement can fall on a boundary point. The subintervals .1 to < .6, .6 to < 1.1, and so on, are convenient and a tally is constructed.

Class i	Class Boundaries	Tally	f_i	Relative frequency, f_i/n
1	0.1 to < 0.6	11111 11111	10	.167
2	0.6 to < 1.1	11111 11111 11111	15	.250
3	1.1 to < 1.6	11111 11111 11111	15	.250
4	1.6 to < 2.1	11111 11111	10	.167
5	2.1 to < 2.6	1111	4	.067
6	2.6 to < 3.1	1	1	.017
7	3.1 to < 3.6	11	2	.033
8	3.6 to < 4.1	1	1	.017
9	4.1 to < 4.6	1	1	.017
10	4.6 to < 5.1		0	.000
11	5.1 to < 5.6	1	1	.017

The relative frequency histogram is shown below.

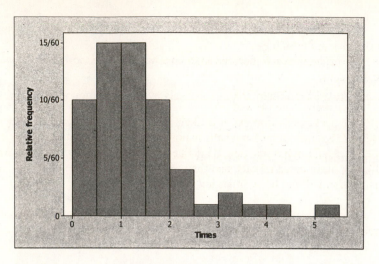

a The distribution is skewed to the right, with several unusually large observations.

b For some reason, one person had to wait 5.2 minutes. Perhaps the supermarket was understaffed that day, or there may have been an unusually large number of customers in the store.

c The two graphs convey the same information. The stem and leaf plot allows us to actually recreate the actual data set, while the histogram does not.

1.35 **a** Histograms will vary from student to student. A typical histogram, generated by *Minitab* is shown on the next page.

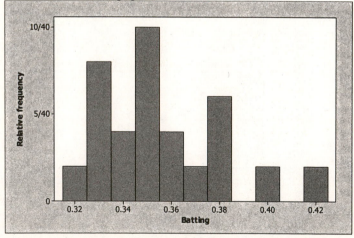

b Since 2 of the 20 players have averages above 0.400, the chance is 2 out of 20 or $2/20 = 0.1$.

1.39 To determine whether a distribution is likely to be skewed, look for the likelihood of observing extremely large or extremely small values of the variable of interest.

a The distribution of non-secured loan sizes might be skewed (a few extremely large loans are possible).
b The distribution of secured loan sizes is not likely to contain unusually large or small values.
c Not likely to be skewed.
d Not likely to be skewed.
e If a package is dropped, it is likely that all the shells will be broken. Hence, a few large number of broken shells is possible. The distribution will be skewed.
f If an animal has one tick, he is likely to have more than one. There will be some "0"s with uninfected rabbits, and then a larger number of large values. The distribution will not be symmetric.

1.43   **a**   Stem and leaf displays may vary from student to student. The most obvious choice is to use the tens digit as the stem and the ones digit as the leaf.

```
 7 | 8 9
 8 | 0 1 7
 9 | 0 1 2 4 4 5 6 6 6 8 8
10 | 1 7 9
11 | 2
```

The display is fairly mound-shaped, with a large peak in the middle.

1.47   Answers will vary from student to student. The students should notice that the distribution is skewed to the right with a few presidents (Truman, Cleveland, and F.D. Roosevelt) casting an unusually large number of vetoes.

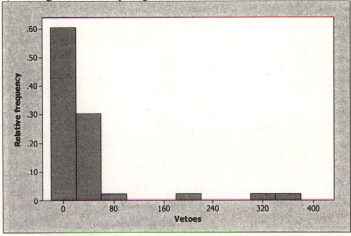

1.51   **a**   The popular vote within each state should vary depending on the size of the state. Since there are several very large states (in population) in the United States, the distribution should be skewed to the right.
b-c Histograms will vary from student to student, but should resemble the histogram generated by *Minitab* in the figure below. The distribution is indeed skewed to the right, with two outliers – California and New York.

5

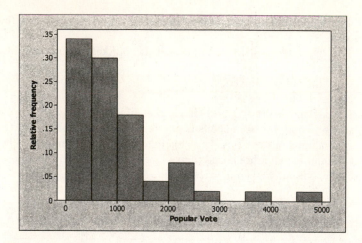

1.55 **a-b** Answers will vary from student to student. The line chart should look similar to the one shown below.

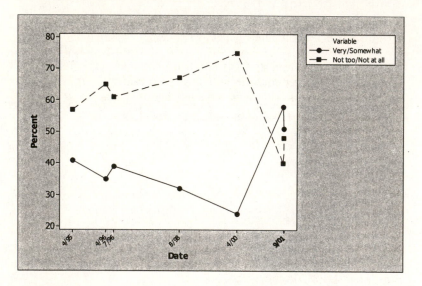

c The percentage of people who were not worried was rising at a slow rate until September 11, 2001, at which time the percentages reversed themselves dramatically. **d** The horizontal axis on the www.gallup.com chart is not an actual time line, so that the time frame in which these changes occur may be distorted.

1.59 **a-b** Answers will vary. A typical histogram is shown below. Notice the gaps and the bimodal nature of the histogram, probably due to the fact that the samples were collected at different locations.

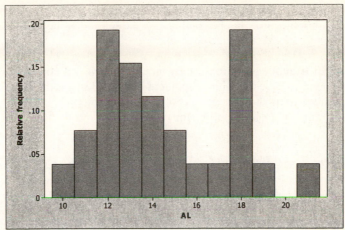

c The dotplot is shown below. The locations are indeed responsible for the unusual gaps and peaks in the relative frequency histogram.

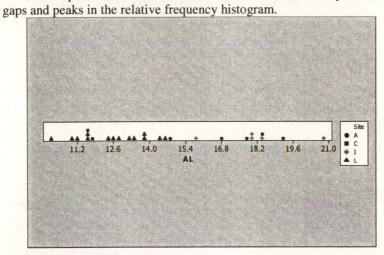

1.63 **a-b** The *Minitab* stem and leaf plot is shown below. The distribution is slightly skewed to the left.

Stem-and-Leaf Display: Percent

```
Stem-and-leaf of Percent   N  = 51
Leaf Unit = 1.0

    1    0  7
    2    0  8
    3    1  0
    4    1  3
    6    1  45
   12    1  666777
   20    1  88888999
  (11)   2  00000001111
   20    2  22222333
   12    2  44444555
    4    2  677
    1    2  9
```

c   Georgia (7.5) and Arkansas (8.0) have gasoline taxes that are somewhat smaller than most, but they may not be "outliers" in the sense that they lie far away from the rest of the measurements in the data set.

1.67   **a-b** The distribution is approximately mound-shaped, with one unusual measurement, in the class with midpoint at 100.8°. Perhaps the person whose temperature was 100.8 has some sort of illness coming on?
c   The value 98.6° is slightly to the right of center.

1.69

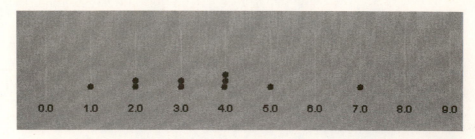

a   The distribution is somewhat mound-shaped (as much as a small set can be); there are no outliers.
b $2/10 = 0.2$

1.73   **a**   There are a few extremely large numbers, indicating that the distribution is probably skewed to the right.
b-c The distribution is indeed skewed right with three possible outliers – Yahoo!, Time Warner and MSN-Microsoft.

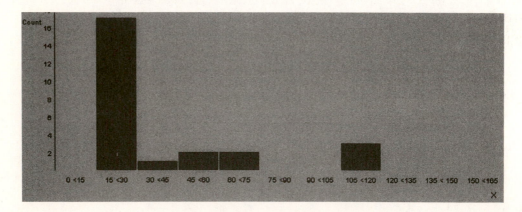

2: Describing Data with Numerical Measures

2.1 **a** The dotplot shown below plots the five measurements along the horizontal axis. Since there are two "1"s, the corresponding dots are placed one above the other. The approximate center of the data appears to be around 1.

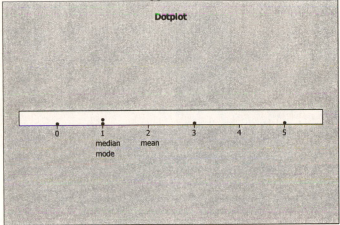

b The mean is the sum of the measurements divided by the number of measurements, or

$$\bar{x} = \frac{\sum x_i}{n} = \frac{0+5+1+1+3}{5} = \frac{10}{5} = 2$$

To calculate the median, the observations are first ranked from smallest to largest: 0, 1, 1, 3, 5. Then since $n = 5$, the position of the median is $0.5(n+1) = 3$, and the median is the 3rd ranked measurement, or $m = 1$. The mode is the measurement occurring most frequently, or mode = 1.

c The three measures in part **b** are located on the dotplot. Since the median and mode are to the left of the mean, we conclude that the measurements are skewed to the right.

2.5 **a** Although there may be a few households who own more than one DVD player, the majority should own either 0 or 1. The distribution should be slightly skewed to the right.

b Since most households will have only one DVD player, we guess that the mode is 1.

c The mean is

$$\bar{x} = \frac{\sum x_i}{n} = \frac{1+0+\cdots+1}{25} = \frac{27}{25} = 1.08$$

To calculate the median, the observations are first ranked from smallest to largest: There are six 0s, thirteen 1s, four 2s, and two 3s. Then since $n = 25$, the position of

the median is $0.5(n+1) = 13$, which is the 13th ranked measurement, or $m = 1$. The mode is the measurement occurring most frequently, or
mode = 1.

d   The relative frequency histogram is shown below, with the three measures superimposed. Notice that the mean falls slightly to the right of the median and mode, indicating that the measurements are slightly skewed to the right.

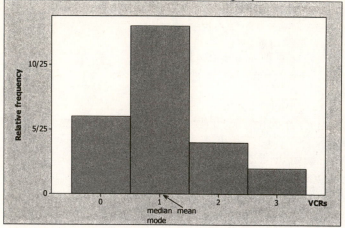

2.9   The distribution of sports salaries will be skewed to the right, because of the very high salaries of some sports figures. Hence, the median salary would be a better measure of center than the mean.

2.13   **a** $\bar{x} = \dfrac{\sum x_i}{n} = \dfrac{12}{5} = 2.4$

b   Create a table of differences, $(x_i - \bar{x})$ and their squares, $(x_i - \bar{x})^2$.

x_i	$x_i - \bar{x}$	$(x_i - \bar{x})^2$
2	−0.4	0.16
1	−1.4	1.96
1	−1.4	1.96
3	0.6	0.36
5	2.6	6.76
Total	0	11.20

Then
$$ s^2 = \frac{\sum (x_i - \bar{x})^2}{n-1} = \frac{(2-2.4)^2 + \cdots + (5-2.4)^2}{4} = \frac{11.20}{4} = 2.8 $$

c   The sample standard deviation is the positive square root of the variance or
$$ s = \sqrt{s^2} = \sqrt{2.8} = 1.673 $$

d   Calculate $\sum x_i^2 = 2^2 + 1^2 + \cdots + 5^2 = 40$. Then

10

$$s^2 = \frac{\sum x_i^2 - \frac{\left(\sum x_i\right)^2}{n}}{n-1} = \frac{40 - \frac{(12)^2}{5}}{4} = \frac{11.2}{4} = 2.8 \text{ and } s = \sqrt{s^2} = \sqrt{2.8} = 1.673.$$

The results of parts **a** and **b** are identical.

2.17 **a** The range is $R = 2.39 - 1.28 = 1.11$.

b Calculate $\sum x_i^2 = 1.28^2 + 2.39^2 + \cdots + 1.51^2 = 15.415$. Then

$$s^2 = \frac{\sum x_i^2 - \frac{\left(\sum x_i\right)^2}{n}}{n-1} = \frac{15.451 - \frac{(8.56)^2}{5}}{4} = \frac{.76028}{4} = .19007$$

and $s = \sqrt{s^2} = \sqrt{.19007} = .436$

c The range, $R = 1.11$, is $1.11/.436 = 2.5$ standard deviations.

2.19 **a** The range of the data is $R = 6 - 1 = 5$ and the range approximation with $n = 10$ is

$$s \approx \frac{R}{3} = 1.67$$

b The standard deviation of the sample is

$$s = \sqrt{s^2} = \sqrt{\frac{\sum x_i^2 - \frac{\left(\sum x_i\right)^2}{n}}{n-1}} = \sqrt{\frac{130 - \frac{(32)^2}{10}}{9}} = \sqrt{3.0667} = 1.751$$

which is very close to the estimate for part **a**.

c-e From the dotplot on the next page, you can see that the data set is not mound-shaped. Hence you can use Tchebysheff's Theorem, but not the Empirical Rule to describe the data.

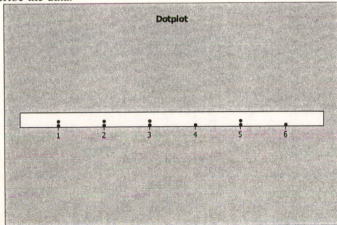

Dotplot

2.21 **a** The interval from 40 to 60 represents $\mu \pm \sigma = 50 \pm 10$. Since the distribution is relatively mound-shaped, the proportion of measurements between 40 and 60 is 68% according to the Empirical Rule and is shown on the next page

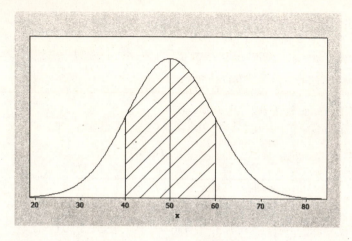

b   Again, using the Empirical Rule, the interval $\mu \pm 2\sigma = 50 \pm 2(10)$ or between 30 and 70 contains approximately 95% of the measurements.

c   Refer to the figure below.

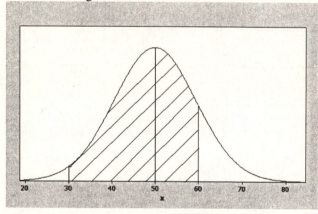

Since approximately 68% of the measurements are between 40 and 60, the symmetry of the distribution implies that 34% of the measurements are between 50 and 60. Similarly, since 95% of the measurements are between 30 and 70, approximately 47.5% are between 30 and 50. Thus, the proportion of measurements between 30 and 60 is

$$0.34 + 0.475 = 0.815$$

d   From the figure in part **a**, the proportion of the measurements between 50 and 60 is 0.34 and the proportion of the measurements which are greater than 50 is 0.50. Therefore, the proportion that are greater than 60 must be

$$0.5 - 0.34 = 0.16$$

2.25   According to the Empirical Rule, if a distribution of measurements is approximately mound-shaped,

a   approximately 68% or 0.68 of the measurements fall in the interval
$\mu \pm \sigma = 12 \pm 2.3$ or 9.7 to 14.3

b   approximately 95% or 0.95 of the measurements fall in the interval
$\mu \pm 2\sigma = 12 \pm 4.6$ or 7.4 to 16.6

c   approximately 99.7% or 0.997 of the measurements fall in the interval
$\mu \pm 3\sigma = 12 \pm 6.9$ or 5.1 to 18.9

Therefore, approximately 0.3% or 0.003 will fall outside this interval.

2.31   **a**   We choose to use 12 classes of length 1.0. The tally and the relative frequency histogram follow.

Class i	Class Boundaries	Tally	f_i	Relative frequency, f_i/n
1	2 to < 3	1	1	1/70
2	3 to < 4	1	1	1/70
3	4 to < 5	111	3	3/70
4	5 to < 6	11111	5	5/70
5	6 to < 7	11111	5	5/70
6	7 to < 8	11111 11111 11	12	12/70
7	8 to < 9	11111 11111 11111 111	18	18/70
8	9 to < 10	11111 11111 11111	15	15/70
9	10 to < 11	11111 1	6	6/70
10	11 to < 12	111	3	3/70
11	12 to < 13		0	0
12	13 to < 14	1	1	1/70

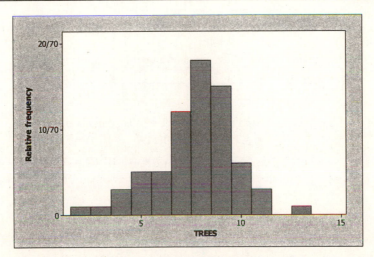

b   Calculate $n = 70$, $\sum x_i = 541$ and $\sum x_i^2 = 4453$. Then $\bar{x} = \dfrac{\sum x_i}{n} = \dfrac{541}{70} = 7.729$ is an estimate of μ.

c   The sample standard deviation is

13

$$s = \sqrt{\frac{\sum x_i^2 - \frac{(\sum x_i)^2}{n}}{n-1}} = \sqrt{\frac{4453 - \frac{(541)^2}{70}}{69}} = \sqrt{3.9398} = 1.985$$

The three intervals, $\bar{x} \pm ks$ for $k = 1, 2, 3$ are calculated below. The table shows the actual percentage of measurements falling in a particular interval as well as the percentage predicted by Tchebysheff's Theorem and the Empirical Rule. Note that the Empirical Rule should be fairly accurate, as indicated by the mound-shape of the histogram in part **a**.

k	$\bar{x} \pm ks$	Interval	Fraction in Interval	Tchebysheff	Empirical Rule
1	7.729 ± 1.985	5.744 to 9.714	$50/70 = 0.71$	at least 0	≈ 0.68
2	7.729 ± 3.970	3.759 to 11.699	$67/70 = 0.96$	at least 0.75	≈ 0.95
3	7.729 ± 5.955	1.774 to 13.684	$70/70 = 1.00$	at least 0.89	≈ 0.997

2.35   **a**   Calculate $R = 2.39 - 1.28 = 1.11$ so that $s \approx R/2.5 = 1.11/2.5 = .444$.

b   In Exercise 2.17, we calculated $\sum x_i = 8.56$ and

$\sum x_i^2 = 1.28^2 + 2.39^2 + \cdots + 1.51^2 = 15.415$. Then

$$s^2 = \frac{\sum x_i^2 - \frac{(\sum x_i)^2}{n}}{n-1} = \frac{15.451 - \frac{(8.56)^2}{5}}{4} = \frac{.76028}{4} = .19007$$

and $s = \sqrt{s^2} = \sqrt{.19007} = .436$, which is very close to our estimate in part **a**.

2.39   **a**   The data in this exercise have been arranged in a frequency table.

x_i	0	1	2	3	4	5	6	7	8	9	10
f_i	10	5	3	2	1	1	1	0	0	1	1

Using the frequency table and the grouped formulas, calculate

$$\sum x_i f_i = 0(10) + 1(5) + \cdots + 10(1) = 51$$
$$\sum x_i^2 f_i = 0^2(10) + 1^2(5) + \cdots + 10^2(1) = 293$$

Then

$$\bar{x} = \frac{\sum x_i f_i}{n} = \frac{51}{25} = 2.04$$

$$s^2 = \frac{\sum x_i^2 f_i - \frac{(\sum x_i f_i)^2}{n}}{n-1} = \frac{293 - \frac{(51)^2}{25}}{24} = 7.873 \text{ and}$$

$s = \sqrt{7.873} = 2.806$.

b-c   The three intervals $\bar{x} \pm ks$ for $k = 1, 2, 3$ are calculated in the table along with the actual proportion of measurements falling in the intervals. Tchebysheff's Theorem is satisfied and the approximation given by the Empirical Rule are fairly close for $k = 2$ and $k = 3$.

k	$\bar{x} \pm ks$	Interval	Fraction in Interval	Tchebysheff	Empirical Rule
1	2.04 ± 2.806	-0.766 to 4.846	$21/25 = 0.84$	at least 0	≈ 0.68
2	2.04 ± 5.612	-3.572 to 7.652	$23/25 = 0.92$	at least 0.75	≈ 0.95
3	2.04 ± 8.418	-6.378 to 10.458	$25/25 = 1.00$	at least 0.89	≈ 0.997

2.41   The data have already been sorted. Find the positions of the quartiles, and the measurements that are just above and below those positions. Then find the quartiles by interpolation.

Sorted Data Set	Position of Q_1	Above and below	Q_1
1, 1.5, 2, 2, 2.2	$.25(6) = 1.5$	1 and 1.5	1.25
0, 1.7, 1.8, 3.1, 3.2, 7, 8, 8.8, 8.9, 9, 10	$.25(12) = 3$	None	1.8
.23, .30, .35, .41, .56, .58, .76, .80	$.25(9) = 2.25$	.30 and .35	$.30 + .25(.05) = .3125$

Position of Q_3	Above and below	Q_3
$.75(6) = 4.5$	2 and 2.2	2.1
$.75(12) = 9$	None	8.9
$.75(9) = 6.75$	.58 and .76	$.58 + .75(.18) = .7150$

2.45   The ordered data are:

$$2, 3, 4, 5, 6, 6, 6, 7, 8, 9, 9, 10, 22$$

For $n = 13$, the position of the median is $0.5(n+1) = 0.5(13+1) = 7$ and $m = 6$. The positions of the quartiles are $0.25(n+1) = 3.5$ and $0.75(n+1) = 10.5$, so that $Q_1 = 4.5$, $Q_3 = 9$, and $IQR = 9 - 4.5 = 4.5$.

The *lower and upper fences* are:

$$Q_1 - 1.5IQR = 4.5 - 6.75 = -2.25$$
$$Q_3 + 1.5IQR = 9 + 6.75 = 15.75$$

The value $x = 22$ lies outside the upper fence and is an outlier. The box plot is shown below. The lower whisker connects the box to the smallest value that is not an outlier, which happens to be the minimum value, $x = 2$. The upper whisker connects the box to the largest value that is not an outlier or $x = 10$.

15

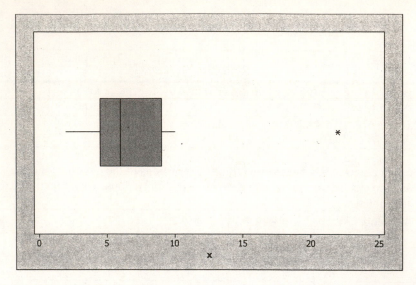

2.49   **a**   For $n = 18$, the position of the median is $0.5(n+1) = 9.5$ and the positions of the quartiles are $0.25(n+1) = 4.75$ and $0.75(n+1) = 14.25$. The lower quartile is ¼ the way between the 4th and 5th measurements and the upper quartile is ¾ the way between the 14th and 15th measurements. The sorted measurements are shown below.

Favre: 10, 12, 13, 14, 15, 15, 18, 19, 21, 22, 22, 23, 23, 23, 23, 25, 25, 26
McNabb: 9, 10, 11, 15, 15, 16, 16, 17, 18, 18, 18, 18, 19, 21, 21, 23, 24, 27

For Brett Favre, $m = (21+22)/2 = 21.5$, $Q_1 = 14 + 0.75(15-14) = 14.75$ and $Q_3 = 23 + 0.25(23-23) = 23$.

For Donovan McNabb, $m = (18+18)/2 = 18$, $Q_1 = 15 + 0.75(15-15) = 15$ and $Q_3 = 21 + 0.25(21-21) = 21$.

Then the five-number summaries are

	Min	Q_1	Median	Q_3	Max
Favre	10	14.75	21.5	23	26
McNabb	9	15	18	21	27

b   For Brett Favre, calculate $IQR = Q_3 - Q_1 = 23 - 14.75 = 8.25$. Then the *lower and upper fences* are:

$$Q_1 - 1.5IQR = 14.75 - 12.375 = 2.375$$
$$Q_3 + 1.5IQR = 23 + 12.375 = 35.375$$

For Donovan McNabb, calculate $IQR = Q_3 - Q_1 = 21 - 15 = 6$. Then the *lower and upper fences* are:

$$Q_1 - 1.5IQR = 15 - 9 = 6$$
$$Q_3 + 1.5IQR = 21 + 9 = 30$$

There are no outliers, and the box plots are shown on the next page.

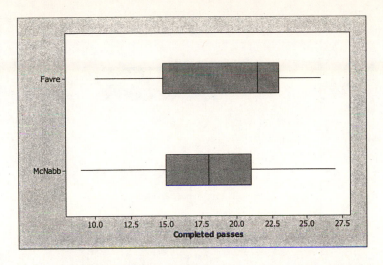

c   Answers will vary. The Favre distribution is skewed left, while the Donovan distribution is roughly symmetric, probably mound-shaped. The McNabb distribution is slightly more variable; Favre has a higher median number of completed passes.

2.53   Answers will vary. The student should notice the outliers in the female group, that the median female temperature is higher than the median male temperature.

2.55   **a**   The ordered sets are shown below:

	Generic					**Sunmaid**			
24	25	25	25	26		22	24	24	24	24
26	26	26	26	27		25	25	27	28	28
27	28	28	28			28	28	29	30	

For $n = 14$, the position of the median is $0.5(n+1) = 0.5(14+1) = 7.5$ and the positions of the quartiles are $0.25(n+1) = 3.75$ and $0.75(n+1) = 11.25$, so that

> **Generic:** $m = 26, Q_1 = 25, Q_3 = 27.25,$ and $IQR = 27.25 - 25 = 2.25$
>
> **Sunmaid:** $m = 26, Q_1 = 24, Q_3 = 28,$ and $IQR = 28 - 24 = 4$

b   **Generic:**   *Lower and upper fences* are:
$$Q_1 - 1.5IQR = 25 - 3.375 = 21.625$$
$$Q_3 + 1.5IQR = 27.25 + 3.375 = 30.625$$

Sunmaid: *Lower and upper fences* are:
$$Q_1 - 1.5IQR = 24 - 6 = 18$$
$$Q_3 + 1.5IQR = 28 + 6 = 34$$

The box plots are shown on the next page. There are no outliers.

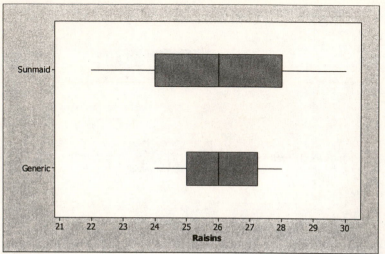

d If the boxes are not being underfilled, the average size of the raisins is roughly the same for the two brands. However, since the number of raisins is more variable for the Sunmaid brand, it would appear that some of the Sunmaid raisins are large while others are small. The individual sizes of the generic raisins are not as variable.

2.59 The ordered data are shown below.

0.2	2.0	4.3	8.2	14.7
0.2	2.1	4.4	8.3	16.7
0.3	2.4	5.6	8.7	18.0
0.4	2.4	5.8	9.0	18.0
1.0	2.7	6.1	9.6	18.4
1.2	3.3	6.6	9.9	19.2
1.3	3.5	6.9	11.4	23.1
1.4	3.7	7.4	12.6	24.0
1.6	3.9	7.4	13.5	26.7
1.6	4.1	8.2	14.1	32.3

Since $n = 50$, the position of the median is $0.5(n+1) = 25.5$ and the positions of the lower and upper quartiles are $0.25(n+1) = 12.75$ and $0.75(n+1) = 38.25$.

Then $m = (6.1+6.6)/2 = 6.35$, $Q_1 = 2.1+0.75(2.4-2.1) = 2.325$ and $Q_3 = 12.6+0.25(13.5-12.6) = 12.825$. Then $IQR = 12.825-2.325 = 10.5$.

The *lower and upper fences* are:
$$Q_1 - 1.5IQR = 2.325 - 15.75 = -13.425$$
$$Q_3 + 1.5IQR = 12.825 + 15.75 = 28.575$$

and the box plot is shown on the next page. There is one outlier, $x = 32.3$. The distribution is skewed to the right.

18

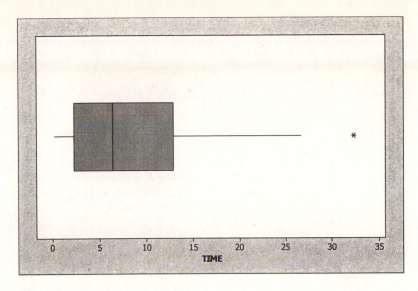

2.63   The following information is available:

$$n = 400, \ \overline{x} = 600, \ s^2 = 4900$$

The standard deviation of these scores is then 70, and the results of Tchebysheff's Theorem follow:

k	$\overline{x} \pm ks$	Interval	Tchebysheff
1	600 ± 70	530 to 670	at least 0
2	600 ± 140	460 to 740	at least 0.75
3	600 ± 210	390 to 810	at least 0.89

If the distribution of scores is mound-shaped, we use the Empirical Rule, and conclude that approximately 68% of the scores would lie in the interval 530 to 670 (which is $\overline{x} \pm s$). Approximately 95% of the scores would lie in the interval 460 to 740.

2.69   If the distribution is mound-shaped, then almost all of the measurements will fall in the interval $\mu \pm 3\sigma$, which is an interval 6σ in length. That is, the range of the measurements should be approximately 6σ. In this case, the range is $800 - 200 = 600$, so that $\sigma \approx 600/6 = 100$.

2.73   The diameters of the trees are approximately mound-shaped with mean 14 and standard deviation 2.8.
a   The value $x = 8.4$ lies two standard deviations below the mean, while the value $x = 22.4$ is three standard deviations above the mean. Use the Empirical Rule. The fraction of trees with diameters between 8.4 and 14 is half of 0.95 or 0.475, while the fraction of trees with diameters between 14 and 22.4 is half of 0.997 or 0.4985. The total fraction of trees with diameters between 8.4 and 22.4 is

$$0.475 + 0.4985 = .9735$$

19

b   The value $x = 16.8$ lies one standard deviation above the mean. Using the Empirical Rule, the fraction of trees with diameters between 14 and 16.8 is half of 0.68 or 0.34, and the fraction of trees with diameters greater than 16.8 is

$$0.5 - 0.34 = .16$$

2.77   **a**   The percentage of colleges that have between 145 and 205 teachers corresponds to the fraction of measurements expected to lie within two standard deviations of the mean. Tchebysheff's Theorem states that this fraction will be at least ¾ or 75%.

       b   If the population is normally distributed, the Empirical Rule is appropriate and the desired fraction is calculated. Referring to the normal distribution shown below, the fraction of area lying between 175 and 190 is 0.34, so that the fraction of colleges having more than 190 teachers is $0.5 - 0.34 = 0.16$.

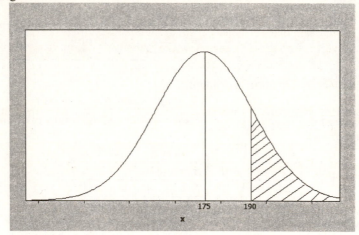

2.81   **a**   Calculate $n = 50$, $\sum x_i = 418$, so that $\bar{x} = \dfrac{\sum x_i}{n} = \dfrac{418}{50} = 8.36$.

       b   The position of the median is $.5(n+1) = 25.5$ and $m = (4 + 4)/2 = 4$.

       c   Since the mean is larger than the median, the distribution is skewed to the right.

       d   Since $n = 50$, the positions of Q_1 and Q_3 are $.25(51) = 12.75$ and $.75(51) = 38.25$, respectively Then $Q_1 = 0 + 0.75(1 - 0) = 12.75$,

$Q_3 = 17 + .25(19 - 17) = 17.5$ and $IQR = 17.5 - .75 = 16.75$.

The *lower and upper fences* are:

$$Q_1 - 1.5IQR = .75 - 25.125 = -24.375$$
$$Q_3 + 1.5IQR = 17.5 + 25.125 = 42.625$$

and the box plot is shown on the next page. There are no outliers and the data is skewed to the right.

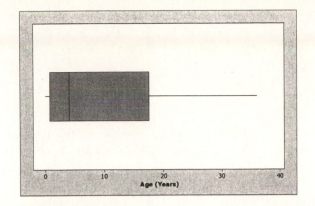

Age (Years)

2.87   **a-b**   As the value of x gets smaller, so does the mean.
c   The median does not change until the green dot is smaller than $x = 10$, at which point the green dot becomes the median.
d   The largest and smallest possible values for the median are $5 \le m \le 10$.

2.91   The box plot shows a distribution that is skewed to the left, but with one outlier to the **right** of the other observations ($x = 520$).

21

3: Describing Bivariate Data

3.1 **a** The side-by-side pie charts are constructed as in Chapter 1 for each of the two groups (men and women) and are displayed below using the percentages shown in the table below.

	Group 1	Group 2	Group 3	Total
Men	23%	31%	46%	100%
Women	8%	57%	35%	100%

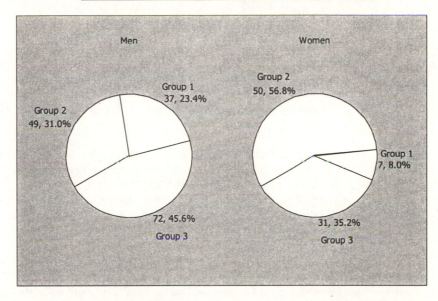

b-c The side-by-side and stacked bar charts in the next two figures measure the frequency of occurrence for each of the three groups. A separate bar (or portion of a bar) is used for men and women.

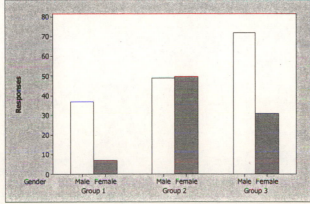

23

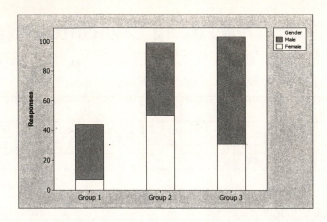

d   The differences in the proportions of men and women in the three groups is most graphically portrayed by the pie charts, since the unequal number of men and women tend to confuse the interpretation of the bar charts. However, the bar charts are useful in retaining the actual frequencies of occurrence in each group, which is lost in the pie chart.

3.5   **a**   The population of interest is the population of responses to the question about free time for all parents and children in the United States. The sample is the set of responses generated for the 198 parents and 200 children in the survey.
b   The data can be considered bivariate if, for each person interviewed, we record the person's relationship (Parent or Child) and their response to the question (just the right amount, not enough, too much, don't know). Since the measurements are not numerical in nature, the variables are qualitative.
c   The entry in a cell represents the number of people who fell into that relationship-opinion category.
d   A pie chart is created for both the "parent" and the "children" categories. The size of each sector angle is proportional to the fraction of measurements falling into that category.

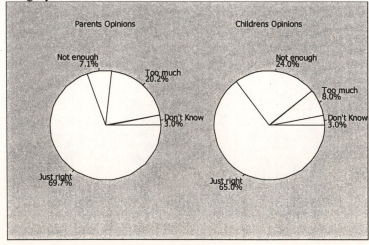

e Either stacked or comparative bar charts could be used, but since the height of the bar represents the frequency of occurrence (and hence is tied to the sample size), this type of chart would be misleading. The comparative pie charts are the best choice.

3.9 Follow the instructions in the My Personal Trainer section. The correct answers are shown in the table.

x	y	xy	Calculate:	Covariance
1	6	6	$n = 3$	$s_{xy} = \dfrac{20 - \dfrac{(6)(12)}{3}}{2} = -2$
3	2	6	$s_x = 1$	
2	4	8	$s_y = 2$	**Correlation Coefficient**
$\sum x = 6$	$\sum y = 12$	$\sum xy = 20$		$r = \dfrac{-2}{1(2)} = -1$

3.13 **a** The scatterplot is shown below.

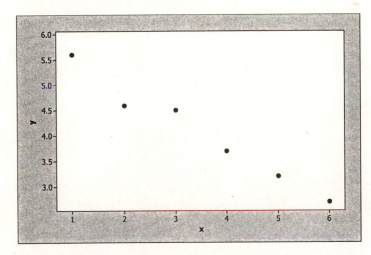

b There appears to be a negative relationship between x and y; that is, as x increase, y decreases.

c Use your scientific calculator to calculate the sums, sums of squares and sum of cross products for the pairs (x_i, y_i).

$$\sum x_i = 21; \ \sum y_i = 24.3; \ \sum x_i^2 = 91; \ \sum y_i^2 = 103.99; \ \sum x_i y_i = 75.3$$

Then the covariance is

$$s_{xy} = \frac{\sum x_i y_i - \dfrac{(\sum x_i)(\sum y_i)}{n}}{n-1} = \frac{75.3 - \dfrac{(21)24.3}{6}}{5} = -1.95$$

25

and the sample standard deviations are

$$s_x = \sqrt{\frac{\sum x_i - \frac{(\sum x_i)^2}{n}}{n-1}} = \frac{91 - \frac{(21)^2}{6}}{5} = 1.8708 \text{ and}$$

$$s_y = \sqrt{\frac{\sum y_i - \frac{(\sum y_i)^2}{n}}{n-1}} = \frac{103.99 - \frac{(24.3)^2}{6}}{5} = 1.0559$$

The correlation coefficient is

$$r = \frac{s_{xy}}{s_x s_y} = \frac{-1.95}{(1.8708)(1.0559)} = -0.987$$

This value of r indicates a strong negative relationship between x and y.

3.17 **a-b** The scatterplot is shown below. There is a slight positive trend between pre- and post-test scores, but the trend is not too pronounced.

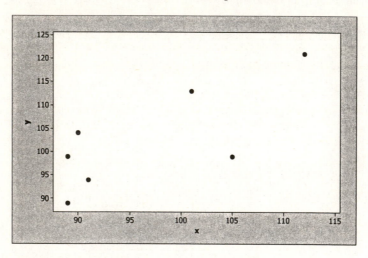

c Calculate
$n = 7;\ \sum x_i = 677;\ \sum y_i = 719;\ \sum x_i^2 = 65,993;\ \sum y_i^2 = 74,585;\ \sum x_i y_i = 70,006$. Then the covariance is

$$s_{xy} = \frac{\sum x_i y_i - \frac{(\sum x_i)(\sum y_i)}{n}}{n-1} = 78.071429$$

The sample standard deviations are $s_x = 9.286447$ and $s_y = 11.056134$ so that $r = 0.760$. This is a relatively strong positive correlation, confirming the interpretation of the scatterplot.

26

3.23   **a**   For each person interviewed in the survey, the following variables are recorded: the service branch (qualitative), the age (quantitative continuous) and the rank (enlisted vs. officer, qualitative).
b   The population of interest is the population of ages for all people in the military. Although the source of this data is not given, it is probably based on census information, in which case the data represents the entire population.
c   A comparative (side-by-side) bar chart has been used. An alternative presentation can be obtained by using comparative pie charts, with the ages divided into eight age groups, and compared for the Army versus the Marine Corps.
d   The enlisted men tend to be younger.
e   The Marine Corps tends to have a much higher percentage of younger enlisted personnel.

3.29   **a-c**   No. There seems to be a large cluster of points in the lower left hand corner showing no apparent relationship between the variables, while 7-10 data points from top left to bottom right show a negative linear relationship.
b   The pattern described in parts **a** and **c** would indicate a weak correlation:

$$r = \frac{s_{xy}}{s_x s_y} = \frac{-83.530}{\sqrt{(712.603)(9346.603)}} = -0.032$$

d   Number of waste sites is only slightly affected by the size of the state. Some other possible explanatory variables might be local environmental regulations, population per square mile, or geographical region in the United States.

3.33   **a**   The 1) number of home networks (quantitative discrete) have been measured, along with the year (quantitative continuous) and the type of network (qualitative).
b-c   Answers will vary. We choose to use a line chart for the two types of networks. As the number of wireless networks increase, the number of wired networks decreases.

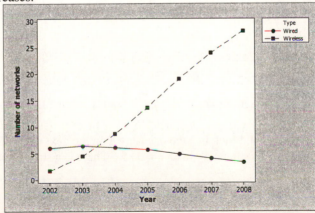

3.37   **a-b**   The scatterplot is shown on the next page. There is a strong positive linear relationship between x and y.

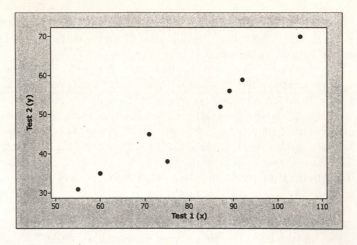

3.38 **a** Calculate

$$n = 8; \; \sum x_i = 634; \; \sum y_i = 386; \; \sum x_i^2 = 52270;$$
$$\sum y_i^2 = 19876; \; \sum x_i y_i = 32136$$

Then the covariance is

$$s_{xy} = \frac{\sum x_i y_i - \dfrac{(\sum x_i)(\sum y_i)}{n}}{n-1} = 220.78571$$

The sample standard deviations are $s_x = 17.010501$ and $s_y = 13.3710775$ so that $r = 0.971$.

b Since the correlation coefficient is so close to 1, the strong correlation indicates that the second and quicker test could be used in place of the longer test-interview.

3.43 **a** Calculate

$$n = 8; \; \sum x_i = 451; \; \sum y_i = 555; \; \sum x_i^2 = 29,619; \; \sum y_i^2 = 43,205; \; \sum x_i y_i = 35,082 \;.$$ Then the covariance is

$$s_{xy} = \frac{\sum x_i y_i - \dfrac{(\sum x_i)(\sum y_i)}{n}}{n-1} = 541.9821$$

The sample standard deviations are $s_x = 24.4770$ and $s_y = 25.9171$ so that $r = 0.8544$.

b-c The scatterplot should look like the one shown on the next page. The correlation coefficient should be close to $r = 0.85$. There is a strong positive trend.

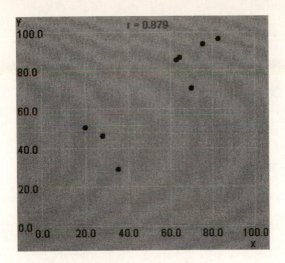

4: Probability and Probability Distributions

4.1   **a**   This experiment involves tossing a single die and observing the outcome. The sample space for this experiment consists of the following simple events:

         E_1: Observe a 1 E_4: Observe a 4
         E_2: Observe a 2 E_5: Observe a 5
         E_3: Observe a 3 E_6: Observe a 6

   b   Events A through F are compound events and are composed in the following manner:

         A: (E_2) D: (E_2)
         B: (E_2, E_4, E_6) E: (E_2, E_4, E_6)
         C: (E_3, E_4, E_5, E_6) F: contains no simple events

   c   Since the simple events E_i, $i = 1, 2, 3, \ldots, 6$ are equally likely, $P(E_i) = 1/6$.

   d   To find the probability of an event, we sum the probabilities assigned to the simple events in that event. For example,

$$P(A) = P(E_2) = \frac{1}{6}$$

Similarly, $P(D) = 1/6; P(B) = P(E) = P(E_2) + P(E_4) + P(E_6) = \dfrac{3}{6} = \dfrac{1}{2};$ and

$P(C) = \dfrac{4}{6} = \dfrac{2}{3}.$ Since event F contains no simple events, $P(F) = 0$.

4.5   **a**   The experiment consists of choosing three coins at random from four. The order in which the coins are drawn is unimportant. Hence, each simple event consists of a triplet, indicating the three coins drawn. Using the letters N, D, Q, and H to represent the nickel, dime, quarter, and half-dollar, respectively, the four possible simple events are listed below.

         E_1: (NDQ) E_2: (NDH) E_3: (NQH) E_4: (DQH)

   b   The event that a half-dollar is chosen is associated with the simple events E_2, E_3, and E_4. Hence,

$$P[\text{choose a half-dollar}] = P(E_2) + P(E_3) + P(E_4) = \frac{1}{4} + \frac{1}{4} + \frac{1}{4} = \frac{3}{4}$$

since each simple event is equally likely.

   c   The simple events along with their monetary values follow:

         E_1 NDQ $0.40
         E_2 NDH   0.65
         E_3 NQH   0.80
         E_4 DQH   0.85

Hence, P[total amount is \$0.60 or more] $= P(E_2) + P(E_3) + P(E_4) = 3/4$.

4.9   The four possible outcomes of the experiment, or simple events, are represented as the cells of a 2×2 table, and have probabilities as given in the table.
a P[adult judged to need glasses] $= .44 + .14 = .58$
b P[adult needs glasses but does not use them] $= .14$
c P[adult uses glasses] $= .44 + .02 = .46$

4.13   **a**   *Experiment*: A taster tastes and ranks three varieties of tea A, B, and C, according to preference.
b   Simple events in S are in triplet form.

$E_1 : (1, 2, 3)$ $E_4 : (2, 3, 1)$

$E_2 : (1, 3, 2)$ $E_5 : (3, 2, 1)$

$E_3 : (2, 1, 3)$ $E_6 : (3, 1, 2)$

Here 1 is assigned to the most desirable, 2 to the next most desirable, and 3 to the least desirable.
c   Define the events   D: variety A is ranked first
                                                          F: variety A is ranked third
Then

$$P(D) = P(E_1) + P(E_2) = 1/6 + 1/6 = 1/3$$

The probability that A is least desirable is
$$P(F) = P(E_5) + P(E_6) = 1/6 + 1/6 = 1/3$$

4.17   Use the *mn* Rule. There are $10(8) = 80$ possible pairs.

4.21   Since order is important, you use *permutations* and $P_5^8 = \dfrac{8!}{3!} = 8(7)(6)(5)(4) = 6720$.

4.25   Since order is unimportant, you use *combinations* and $C_3^{10} = \dfrac{10!}{3!17!} = \dfrac{10(9)(8)}{3(2)(1)} = 120$.

4.29   **a**   Each student has a choice of 52 cards, since the cards are replaced between selections. The *mn* Rule allows you to find the total number of configurations for three students as $52(52)(52) = 140,608$.

b   Now each student must pick a different card. That is, the first student has 52 choices, but the second and third students have only 51 and 50 choices, respectively. The total number of configurations is found using the *mn* Rule or the rule for permutations:

$$mnt = 52(51)(50) = 132,600 \quad \text{or} \quad P_3^{52} = \frac{52!}{49!} = 132,600.$$

c   Let A be the event of interest. Since there are 52 different cards in the deck, there are 52 configurations in which all three students pick the same card (one for each card). That is, there are $n_A = 52$ ways for the event A to occur, out of a total of $N = 140,608$ possible configurations from part **a**. The probability of interest is

$$P(A) = \frac{n_A}{N} = \frac{52}{140,608} = .00037$$

d   Again, let A be the event of interest. There are $n_A = 132,600$ ways (from part **b**) for the event A to occur, out of a total of $N = 140,608$ possible configurations from part **a**, and the probability of interest is

$$P(A) = \frac{n_A}{N} = \frac{132,600}{140,608} = .943$$

4.33   Notice that a sample of 10 nurses will be the same no matter in which order they were selected. Hence, order is unimportant and combinations are used. The number of samples of 10 selected from a total of 90 is

$$C_{10}^{90} = \frac{90!}{10!80!} = \frac{2.0759076\left(10^{19}\right)}{3.6288\left(10^{6}\right)} = 5.720645\left(10^{12}\right)$$

4.37   The situation presented here is analogous to drawing 5 items from a jar (the five members voting in favor of the plaintiff). If the jar contains 5 red and 3 white items (5 women and 3 men), what is the probability that all five items are red? That is, if there is no sex bias, five of the eight members are randomly chosen to be those voting for the plaintiff. What is the probability that all five are women? There are

$$N = C_5^8 = \frac{8!}{5!3!} = 56$$

simple events in the experiment, only one of which results in choosing 5 women. Hence,

$$P(\text{five women}) = \frac{1}{56}.$$

4.41   Follow the instructions given in the My Personal Trainer section. The answers are given in the table.

P(A)	P(B)	Conditions for events A and B	P(A ∩ B)	P(A ∪ B)	P(A|B)
.3	.4	Mutually exclusive	0	.3 + .4 = .7	0
.3	.4	Independent	.3(.4) = .12	.3+.4−(.3)(.4)=.58	.3
.1	.5	Independent	.1(.5) = .05	.1+.5−(.1)(.5)=.55	.1
.2	.5	Mutually exclusive	0	.2 + .5 = .7	0

4.47   Refer to the solution to Exercise 4.1 where the six simple events in the experiment are given, with $P(E_i) = 1/6$.

a $S = \{E_1, E_2, E_3, E_4, E_5, E_6\}$ and $P(S) = 6/6 = 1$

b $P(A|B) = \dfrac{P(A \cap B)}{P(B)} = \dfrac{1/3}{1/3} = 1$

c $B = \{E_1, E_2\}$ and $P(B) = 2/6 = 1/3$

d $A \cap B \cap C$ contains no simple events, and $P(A \cap B \cap C) = 0$

e $P(A \cap B) = P(A|B)P(B) = 1(1/3) = 1/3$

f $A \cap C$ contains no simple events, and $P(A \cap C) = 0$

g $B \cap C$ contains no simple events, and $P(B \cap C) = 0$

h $A \cup C = S$ and $P(A \cup C) = 1$

i $B \cup C = \{E_1, E_2, E_4, E_5, E_6\}$ and $P(B \cup C) = 5/6$

4.48   **a**   From Exercise 4.47, $P(A \cap B) = 1/3$, $P(A|B) = 1$, $P(A) = 1/2$,

$P(A|B) \neq P(A)$, so that A and B are not independent. $P(A \cap B) \neq 0$, so that A and B are not mutually exclusive.

b $P(A|C) = P(A \cap C)/P(C) = 0$, $P(A) = 1/2$, $P(A \cap C) = 0$. Since $P(A|C) \neq P(A)$, A and C are dependent. Since $P(A \cap C) = 0$, A and C are mutually exclusive.

4.49   **a**   Since A and B are independent, $P(A \cap B) = P(A)P(B) = .4(.2) = .08$.

b $P(A \cup B) = P(A) + P(B) - P(A \cap B) = .4 + .2 - (.4)(.2) = .52$

4.55   Define the following events:

   A: project is approved for funding
   D: project is disapproved for funding

For the first group, $P(A_1) = .2$ and $P(D_1) = .8$. For the second group,

$P[\text{same decision as first group}] = .7$ and $P[\text{reversal}] = .3$. That is,

$$P(A_2 \mid A_1) = P(D_2 \mid D_1) = .7 \text{ and } P(A_2 \mid D_1) = P(D_2 \mid A_1) = .3.$$

a $P(A_1 \cap A_2) = P(A_1)P(A_2 \mid A_1) = .2(.7) = .14$

b $P(D_1 \cap D_2) = P(D_1)P(D_2 \mid D_1) = .8(.7) = .56$

c

$$P(D_1 \cap A_2) + P(A_1 \cap D_2) = P(D_1)P(A_2 \mid D_1) + P(A_1)P(D_2 \mid A_1) = .8(.3) + .2(.3) = .30$$

4.59   Fix the birth date of the first person entering the room. Then define the following events:

   A_2: second person's birthday differs from the first
   A_3: third person's birthday differs from the first and second
   A_4: fourth person's birthday differs from all preceding
   $\vdots$
   A_n: n^{th} person's birthday differs from all preceding

34

Then
$$P(A) = P(A_2)P(A_3)\cdots P(A_n) = \left(\frac{364}{365}\right)\left(\frac{363}{365}\right)\cdots\left(\frac{365-n+1}{365}\right)$$

since at each step, one less birth date is available for selection. Since event B is the complement of event A,
$$P(B) = 1 - P(A)$$

a For $n = 3$, $P(A) = \dfrac{(364)(363)}{(365)^2} = .9918$ and $P(B) = 1 - .9918 = .0082$

b For $n = 4$, $P(A) = \dfrac{(364)(363)(362)}{(365)^3} = .9836$ and $P(B) = 1 - .9836 = .0164$

4.63 Define   A: smoke is detected by device A
              B: smoke is detected by device B
If it is given that $P(A) = .95$, $P(B) = .98$, and $P(A \cap B) = .94$.

a $P(A \cup B) = P(A) + P(B) - P(A \cap B) = .95 + .98 - .94 = .99$

b $P(A^C \cap B^C) = 1 - P(A \cup B) = 1 - .99 = .01$

4.69 **a** Use the Law of Total Probability, writing
$$P(A) = P(S_1)P(A \mid S_1) + P(S_2)P(A \mid S_2) = .7(.2) + .3(.3) = .23$$

b Use the results of part a in the form of Bayes' Rule:
$$P(S_i \mid A) = \frac{P(S_i)P(A \mid S_i)}{P(S_1)P(A \mid S_1) + P(S_2)P(A \mid S_2)}$$

For $i = 1$, $P(S_1 \mid A) = \dfrac{.7(.2)}{.7(.2) + .3(.3)} = \dfrac{.14}{.23} = .6087$

For $i = 2$, $P(S_2 \mid A) = \dfrac{.3(.3)}{.7(.2) + .3(.3)} = \dfrac{.09}{.23} = .3913$

4.73 Define   A: machine produces a defective item
              B: worker follows instructions
Then $P(A \mid B) = .01$, $P(B) = .90$, $P(A \mid B^C) = .03$, $P(B^C) = .10$. The probability of interest is
$$P(A) = P(A \cap B) + P(A \cap B^C)$$
$$= P(A \mid B)P(B) + P(A \mid B^C)P(B^C)$$
$$= .01(.90) + .03(.10) = .012$$

4.77 The probability of interest is $P(A \mid H)$ which can be calculated using Bayes' Rule and the probabilities given in the exercise.

$$P(A \mid H) = \frac{P(A)P(H \mid A)}{P(A)P(H \mid A) + P(B)P(H \mid B) + P(C)P(H \mid C)}$$

$$= \frac{.01(.90)}{.01(.90) + .005(.95) + .02(.75)} = \frac{.009}{.02875} = .3130$$

4.83 **a** Since one of the requirements of a probability distribution is that $\sum_x p(x) = 1$, we need

$$p(3) = 1 - (.1 + .3 + .3 + .1) = 1 - .8 = .2$$

b The probability histogram is shown below.

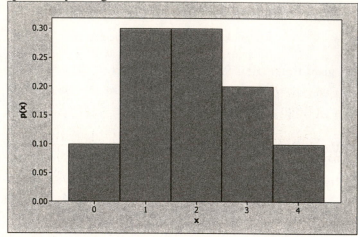

c For the random variable x given here,

$$\mu = E(x) = \sum xp(x) = 0(.1) + 1(.3) + \cdots + 4(.1) = 1.9$$

The variance of x is defined as

$$\sigma^2 = E\left[(x - \mu)^2\right] = \sum (x - \mu)^2 \, p(x) = (0 - 1.9)^2(.1) + (1 - 1.9)^2(.3) + \cdots + (4 - 1.9)^2(.1) = 1.29$$

and $\sigma = \sqrt{1.29} = 1.136$.

d Using the table form of the probability distribution given in the exercise, $P(x > 2) = .2 + .1 = .3$.

e $P(x \le 3) = 1 - P(x = 4) = 1 - .1 = .9$.

4.87 **a-b** On the first try, the probability of selecting the proper key is 1/4. If the key is not found on the first try, the probability changes on the second try. Let F denote a failure to find the key and S denote a success. The random variable is x, the number of keys tried before the correct key is found. The four associated simple events are shown below.

E_1: S $\quad (x = 1)$ $\qquad\qquad$ E_3: FFS $\quad (x = 3)$

E_2: FS $\quad (x = 2)$ $\qquad\qquad$ E_4: FFFS $\quad (x = 4)$

c-d Then

$$p(1) = P(x = 1) = P(S) = 1/4$$

$$p(2) = P(x = 2) = P(FS) = P(F)P(S) = (3/4)(1/3) = 1/4$$

$$p(3) = P(x = 3) = P(FFS) = P(F)P(F)P(S) = (3/4)(2/3)(1/2) = 1/4$$

$$p(4) = P(x = 4) = P(FFFS) = P(F)P(F)P(F)P(S) = (3/4)(2/3)(1/2)(1) = 1/4$$

The probability distribution and probability histogram follow.

x	1	2	3	4
$p(x)$	1/4	1/4	1/4	1/4

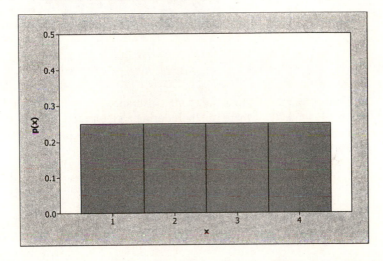

4.91   Let x be the number of drillings until the first success (oil is struck). It is given that the probability of striking oil is $P(O) = .1$, so that the probability of no oil is $P(N) = .9$

a $p(1) = P[\text{oil struck on first drilling}] = P(O) = .1$

$p(2) = P[\text{oil struck on second drilling}]$. This is the probability that oil is not found on the first drilling, but is found on the second drilling. Using the Multiplication Law,

$$p(2) = P(NO) = (.9)(.1) = .09 .$$

Finally, $p(3) = P(NNO) = (.9)(.9)(.1) = .081$.

b-c   For the first success to occur on trial x, $(x - 1)$ failures must occur before the first success. Thus,

$$p(x) = P(NNN \ldots NNO) = (.9)^{x-1}(.1)$$

since there are $(x - 1)$ N's in the sequence. The probability histogram is shown on the next page.

37

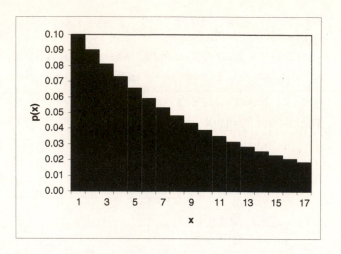

4.95 The random variable G, total gain to the insurance company, will be D if there is no theft, but D – 50,000 if there is a theft during a given year. These two events will occur with probability .99 and .01, respectively. Hence, the probability distribution for G is given below.

G	$p(G)$
D	.99
D – 50,000	.01

The expected gain is

$$E(G) = \sum Gp(G) = .99D + .01(D - 50,000)$$
$$= D - 50,000$$

In order that $E(G) = 1000$, it is necessary to have $1000 = D - 500$ or $D = \$1500$.

4.97 **a** Similar to Exercise 4.91. For the first non-believer to be found on call x, $(x - 1)$ people who *do* believe in heaven must be called before the first non-believer is found. Thus,

$$p(x) = P(NNN \ldots NNY) = (.81)^{x-1}(.19)$$

b As with other phone surveys, there is always a problem of non-response – people who do not answer the telephone or decline to participate in the survey. Also, there is a problem of truthfulness of the response for a question such as this which may be a sensitive subject for some people.

4.101 Define the following events:

A: worker fails to report fraud
B: worker suffers reprisal

It is given that $P(B \mid A^C) = .23$ and $P(A) = .69$. The probability of interest is

$$P(A^C \cap B) = P(B \mid A^C)P(A^C) = .23(.31) = .0713$$

4.105 Two systems are selected from seven, three of which are defective. Denote the seven systems as G_1, G_2, G_3, G_4, D_1, D_2, D_3 according to whether they are good or defective.

38

Each simple event will represent a particular pair of systems chosen for testing, and the sample space, consisting of 21 pairs, is shown below.

$$G_1G_2 \quad G_1D_1 \quad G_2D_3 \quad G_4D_2 \quad G_1G_3 \quad G_1G_2 \quad G_3D_1 \quad G_4D_3$$
$$G_1G_4 \quad G_1D_3 \quad G_3D_2 \quad D_1D_2 \quad G_2G_3 \quad G_2D_1 \quad G_3D_3 \quad D_1D_3$$
$$G_2G_4 \quad G_2D_2 \quad G_4D_1 \quad D_2D_3 \quad G_3G_4$$

Note that the two systems are drawn simultaneously and that order is unimportant in identifying a simple event. Hence, the pairs G_1G_2 and G_2G_1 are not considered to represent two different simple events. The event A, "no defectives are selected", consists of the simple events $G_1G_2, G_1G_3, G_1G_4, G_2G_3, G_2G_4, G_3G_4$. Since the systems are selected at random, any pair has an equal probability of being selected. Hence, the probability assigned to each simple event is 1/21 and $P(A) = 6/21 = 2/7$.

4.109 **a** $P(\text{cold}) = \dfrac{49 + 43 + 34}{276} = \dfrac{126}{276} = .4565$

b Define: F: person has four or five relationships
   S: person has six or more relationships

Then for the two people chosen from the total 276,

$$P(\text{one F and one S}) = P(F \cap S) + P(S \cap F)$$

$$= \left(\frac{100}{276}\right)\left(\frac{96}{275}\right) + \left(\frac{96}{276}\right)\left(\frac{100}{275}\right) = .2530$$

c $P(\text{Three or fewer} \mid \text{cold}) = \dfrac{P(\text{three or fewer} \cap \text{cold})}{P(\text{cold})} = \dfrac{49/276}{126/276} = \dfrac{49}{126} = .3889$

4.113 **a** Define the following events:
   B_1: client buys on first contact
   B_2: client buys on second contact

Since the client may buy on either the first of the second contact, the desired probability is

$$P[\text{client will buy}] = P[\text{client buys on first contact}]$$

$$+ P[\text{client doesn't buy on first, but buys on second}]$$

$$= P(B_1) + (1 - P(B_1))P(B_2) = .4 + (1 - .4)(.55)$$

$$= .73$$

b The probability that the client will not buy is one minus the probability that the client will buy, or $1 - .73 = .27$.

4.117 Each ball can be chosen from the set (4, 6) and there are three such balls. Hence, there are a total of $2(2)(2) = 8$ potential winning numbers.

4.121 **a** Consider a single trial which consists of tossing two coins. A match occurs when either HH or TT is observed. Hence, the probability of a match on a single trial

39

is $P(HH) + P(TT) = 1/4 + 1/4 = 1/2$. Let MMM denote the event "match on trials 1, 2, and 3". Then

$$P(MMM) = P(M)P(M)P(M) = (1/2)^3 = 1/8.$$

b   On a single trial the event A, "two trails are observed" has probability $P(A) = P(TT) = 1/4$. Hence, in three trials

$$P(AAA) = P(A)P(A)P(A) = (1/4)^3 = 1/64$$

c   This low probability would not suggest collusion, since the probability of three matches is low only if we assume that each student is merely guessing at each answer. If the students have studied together or if they both know the correct answer, the probability of a match on a single trial is no longer 1/2, but is substantially higher. Hence, the occurrence of three matches is not unusual.

4.125   Define the events: A: the man waits five minutes or longer
                                      B: the woman waits five minutes or longer
The two events are independent, and $P(A) = P(B) = .2$.

a $P(A^C) = 1 - P(A) = .8$

b $P(A^C B^C) = P(A^C) P(B^C) = (.8)(.8) = .64$

c $P[\text{at least one waits five minutes or longer}]$
$$= 1 - P[\text{neither waits five minutes or longer}] = 1 - P(A^C B^C) = 1 - .64 = .36$$

4.129   Since the first pooled test is positive, we are interested in the probability of requiring five single tests to detect the disease in the single affected person. There are $(5)(4)(3)(2)(1)$ ways of ordering the five tests, and there are $4(3)(2)(1)$ ways of ordering the tests so that the diseased person is given the final test. Hence, the desired probability is $\dfrac{4!}{5!} = \dfrac{1}{5}$.

If two people are diseased, six tests are needed if the last two tests are given to the diseased people. There are $3(2)(1)$ ways of ordering the tests of the other three people and $2(1)$ ways of ordering the tests of the two diseased people. Hence, the probability that six tests will be needed is $\dfrac{2!3!}{5!} = \dfrac{1}{10}$.

4.133   **a**   Define P: shopper prefers Pepsi and C: shopper prefers Coke. Then if there is actually no difference in the taste, P(P) = P(C) = 1/2 and

$$P(\text{all four prefer Pepsi}) = P(PPPP) = [P(P)]^4 = \left(\frac{1}{2}\right)^4 = \frac{1}{16} = .0625$$

b
$P(\text{exactly one prefers Pepsi}) = P(PCCC) + P(CPCC) + P(CCPC) + P(CCCP)$

$$= 4P(P)[P(C)]^3 = 4\left(\frac{1}{2}\right)\left(\frac{1}{2}\right)^3 = \frac{4}{16} = .25$$

40

4.137 Refer to the **Tossing Dice** applet, in which the simple events for this experiment are displayed. Each simple event has a particular value of T associated with it, and by summing the probabilities of all simple events producing a particular value of T, the following probability distribution is obtained. The distribution is mound-shaped.

a-b

T	p(T)	T	p(T)
2	1/36	8	5/36
3	2/36	9	4/36
4	3/36	10	3/36
5	4/36	11	2/36
6	5/36	12	1/36
7	6/36		

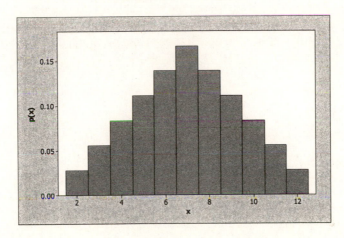

5: Several Useful Discrete Distributions

5.1   Follow the instructions in the Personal Trainer section. The answers are shown in the tables below.

k	0	1	2	3	4	5	6	7	8
$P(x \le k)$	.000	.001	.011	.058	.194	.448	.745	.942	1.000

The Problem	List the Values of x	Write the probability	Rewrite the probability	Find the probability
Three or less	0, 1, 2, 3	$P(x \le 3)$		.058
Three or more	3, 4, 5, 6, 7, 8	$P(x \ge 3)$	$1 - P(x \le 2)$	$1 - .011 = .989$
More than three	4, 5, 6, 7, 8	$P(x > 3)$	$1 - P(x \le 3)$	$1 - .058 = .942$
Fewer than three	0, 1, 2	$P(x < 3)$	$P(x \le 2)$	.011
Between 3 and 5 (inclusive)	3, 4, 5	$P(3 \le x \le 5)$	$P(x \le 5) - P(x \le 2)$	$.448 - .011 = .437$
Exactly three	3	$P(x = 3)$	$P(x \le 3) - P(x \le 2)$	$.058 - .011 = .047$

5.5   **a** $C_2^8 (.3)^2 (.7)^6 = \dfrac{8(7)}{2(1)}(.09)(.117649) = .2965$

      b $C_0^4 (.05)^0 (.95)^4 = (.95)^4 = .8145$

      c $C_3^{10} (.5)^3 (.5)^7 = \dfrac{10(9)(8)}{3(2)(1)}(.5)^{10} = .1172$

      d $C_1^7 (.2)^1 (.8)^6 = 7(.2)(.8)^6 = .3670$

5.11   **a**   For $n = 10$ and $p = .4$, $P(x = 4) = C_4^{10}(.4)^4 (.6)^6 = .251$.

      b   To calculate $P(x \ge 4) = p(4) + p(5) + \cdots + p(10)$ it is easiest to write
$$P(x \ge 4) = 1 - P(x < 4) = 1 - P(x \le 3).$$

These probabilities can be found individually using the binomial formula, or alternatively using the cumulative binomial tables in Appendix I.

$$P(x = 0) = C_0^{10}(.4)^0 (.6)^{10} = .006 \qquad P(x = 1) = C_1^{10}(.4)^1 (.6)^9 = .040$$

$$P(x = 2) = C_2^{10}(.4)^2 (.6)^8 = .121 \qquad P(x = 3) = C_3^{10}(.4)^3 (.6)^7 = .215$$

The sum of these probabilities gives $P(x \le 3) = .382$ and $P(x \ge 4) = 1 - .382 = .618$.

      c   Use the results of parts **a** and **b**.
$$P(x > 4) = 1 - P(x \le 4) = 1 - (.382 + .251) = .367$$

d   From part **c**, $P(x \leq 4) = P(x \leq 3) + P(x = 4) = .382 + .251 = .633$.

e $\mu = np = 10(.4) = 4$

f $\sigma = \sqrt{npq} = \sqrt{10(.4)(.6)} = \sqrt{2.4} = 1.549$

5.15   **a** $P[x < 12] = P[x \leq 11] = .748$

        b $P[x \leq 6] = .610$

        c $P[x > 4] = 1 - P[x \leq 4] = 1 - .633 = .367$

        d $P[x \geq 6] = 1 - P[x \leq 5] = 1 - .034 = .966$

        e $P[3 < x < 7] = P[x \leq 6] - P[x \leq 3] = .828 - .172 = .656$

5.19   **a** $p(0) = C_0^{20}(.1)^0 (.9)^{20} = .1215767$ $p(3) = C_3^{20}(.1)^3 (.9)^{17} = .1901199$

             $p(1) = C_1^{20}(.1)^1 (.9)^{19} = .2701703$ $p(4) = C_4^{20}(.1)^4 (.9)^{16} = .0897788$

             $p(2) = C_2^{20}(.1)^2 (.9)^{18} = .2851798$

so that $P[x \leq 4] = p(0) + p(1) + p(2) + p(3) + p(4) = .9568255$

        b   Using Table 1, Appendix I, $P[x \leq 4]$ is read directly as .957.

        c   Adding the entries for $x = 0,1,2,3,4$, we have $P[x \leq 4] = .956826$.

        d $\mu = np = 20(.1) = 2$ and $\sigma = \sqrt{npq} = \sqrt{1.8} = 1.3416$

        e   For $k = 1$, $\mu \pm \sigma = 2 \pm 1.342$ or .658 to 3.342 so that

$$P[.658 \leq x \leq 3.342] = P[1 \leq x \leq 3] = .2702 + .2852 + .1901 = .7455$$

             For $k = 2$, $\mu \pm 2\sigma = 2 \pm 2.683$ or $-.683$ to 4.683 so that

$$P[-.683 \leq x \leq 4.683] = P[0 \leq x \leq 4] = .9569$$

             For $k = 3$, $\mu \pm 3\sigma = 2 \pm 4.025$ or -2.025 to 6.025 so that

$$P[-2.025 \leq x \leq 6.025] = P[0 \leq x \leq 6] = .9977$$

        f   The results are consistent with Tchebysheff's Theorem and the Empirical Rule.

5.23   Define x to be the number of alarm systems that are triggered. Then $p = P[\text{alarm is triggered}] = .99$ and $n = 9$. Since there is a table available in Appendix I for $n = 9$ and $p = .99$, you should use it rather than the binomial formula to calculate the necessary probabilities.

        a $P[\text{at least one alarm is triggered}] = P(x \geq 1) = 1 - P(x = 0) = 1 - .000 = 1.000$.

        b $P[\text{more than seven}] = P(x > 7) = 1 - P(x \leq 7) = 1 - .003 = .997$

        c $P[\text{eight or fewer}] = P(x \leq 8) = .086$

5.25   Define x to be the number of cars that are black. Then $p = P[\text{black}] = .1$ and $n = 25$. Use Table 1 in Appendix I.

a $P(x \geq 5) = 1 - P(x \leq 4) = 1 - .902 = .098$

b $P(x \leq 6) = .991$

c $P(x > 4) = 1 - P(x \leq 4) = 1 - .902 = .098$

d $P(x = 4) = P(x \leq 4) - P(x \leq 3) = .902 - .764 = .138$

e $P(3 \leq x \leq 5) = P(x \leq 5) - P(x \leq 2) = .967 - .537 = .430$

f $P(\text{more than 20 } not \text{ black}) = P(\text{less than 5 black}) = P(x \leq 4) = .902$

5.29   Define x to be the number of fields infested with whitefly.
Then $p = P[\text{infected field}] = .1$ and $n = 100$.

    a $\mu = np = 100(.1) = 10$

    b   Since n is large, this binomial distribution should be fairly mound-shaped, even
though $p = .1$. Hence you would expect approximately 95% of the measurements to
lie within two standard deviation of the mean with $\sigma = \sqrt{npq} = \sqrt{100(.1)(.9)} = 3$. The
limits are calculated as

$$\mu \pm 2\sigma \Rightarrow 10 \pm 6 \text{ or from 4 to 16}$$

    c   From part **b**, a value of $x = 25$ would be very unlikely, assuming that the
characteristics of the binomial experiment are met and that $p = .1$. If this value were
actually observed, it might be possible that the trials (fields) are not independent.
This could easily be the case, since an infestation in one field might quickly spread to
a neighboring field. This is evidence of *contagion*.

5.33   Define x to be the number of Americans who are "tasters". Then, $n = 20$ and $p = .7$.
Using the binomial tables in Appendix I,

    a $P(x \geq 17) = 1 - P(x \leq 16) = 1 - .893 = .107$

    b $P(x \leq 15) = .762$

5.35   Follow the instructions in the My Personal Trainer section. The answers are shown in
the table below.

Probability	Formula	Calculated value
$P(x = 0)$	$\dfrac{2.5^0 e^{-2.5}}{0!}$	.0821
$P(x = 1)$	$\dfrac{2.5^1 e^{-2.5}}{1!}$	.2052
$P(x = 2)$	$\dfrac{2.5^2 e^{-2.5}}{2!}$	.2565
$P(2 \text{ or fewer successes})$	$P(x = 0) + P(x = 1) + P(x = 2)$	.5438

5.39 Using $p(x) = \dfrac{\mu^x e^{-\mu}}{x!} = \dfrac{2^x e^{-2}}{x!}$,

    a $\quad P[x=0] = \dfrac{2^0 e^{-2}}{0!} = .135335$

    b $\quad P[x=1] = \dfrac{2^1 e^{-2}}{1!} = .27067$

    c $\quad P[x>1] = 1 - P[x \le 1] = 1 - .135335 - .27067 = .593994$

    d $\quad P[x=5] = \dfrac{2^5 e^{-2}}{5!} = .036089$

5.43 Let x be the number of misses during a given month. Then x has a Poisson distribution with $\mu = 5$.

    a $\quad p(0) = e^{-5} = .0067$ **b** $\quad p(5) = \dfrac{5^5 e^{-5}}{5!} = .1755$

    c $\quad P[x \ge 5] = 1 - P[x \le 4] = 1 - .440 = .560$ from Table 2.

5.47 The random variable x, number of bacteria, has a Poisson distribution with $\mu = 2$. The probability of interest is
$$P[x \text{ exceeds maximum count}] = P[x > 5]$$
Using the fact that $\mu = 2$ and $\sigma = 1.414$ from Exercise 5.47, most of the observations should fall within $\mu \pm 2\sigma$ or 0 to 4. Hence, it is unlikely that x will exceed 5. In fact, the exact Poisson probability is $P[x > 5] = .017$.

5.51 The formula for $p(x)$ is $p(x) = \dfrac{C_x^4 C_{3-x}^{11}}{C_3^{15}}$ for $x = 0, 1, 2, 3$

    a $\quad p(0) = \dfrac{C_0^4 C_3^{11}}{C_3^{15}} = \dfrac{165}{455} = .36 \qquad\qquad p(1) = \dfrac{C_1^4 C_2^{11}}{C_3^{15}} = \dfrac{220}{455} = .48$

           $p(2) = \dfrac{C_2^4 C_1^{11}}{C_3^{15}} = \dfrac{66}{455} = .15 \qquad\qquad p(3) = \dfrac{C_3^4 C_0^{11}}{C_3^{15}} = \dfrac{4}{455} = .01$

    b   The probability histogram is shown below.

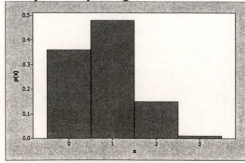

c   Using the formulas given in Section 5.4.

$$\mu = E(x) = n\left(\frac{M}{N}\right) = 3\left(\frac{4}{15}\right) = .8$$

$$\sigma^2 = n\left(\frac{M}{N}\right)\left(\frac{N-M}{N}\right)\left(\frac{N-n}{N-1}\right) = 3\left(\frac{4}{15}\right)\left(\frac{15-4}{15}\right)\left(\frac{15-3}{15-1}\right) = .50286$$

d   Calculate the intervals

$$\mu \pm 2\sigma = .8 \pm 2\sqrt{.50286} = .8 \pm 1.418 \text{ or } -.618 \text{ to } 2.218$$

$$\mu \pm 3\sigma = .8 \pm 3\sqrt{.50286} = .8 \pm 1.418 \text{ or } -1.327 \text{ to } 2.927$$

Then,

$$P[-.618 \le x \le 2.218] = p(0) + p(1) + p(2) = .99$$

$$P[-1.327 \le x \le 2.927] = p(0) + p(1) + p(2) = .99$$

These results agree with Tchebysheff's Theorem.

5.55   **a**   The random variable x has a hypergeometric distribution with $N = 8, M = 5$ and $n = 3$. Then

$$p(x) = \frac{C_x^5 C_{3-x}^3}{C_3^8} \text{ for } x = 0, 1, 2, 3$$

b $P(x = 3) = \dfrac{C_3^5 C_0^3}{C_3^8} = \dfrac{10}{56} = .1786$ **c** $P(x = 0) = \dfrac{C_0^5 C_3^3}{C_3^8} = \dfrac{1}{56} = .01786$

d $P(x \le 1) = \dfrac{C_0^5 C_3^3}{C_3^8} + \dfrac{C_1^5 C_2^3}{C_3^8} = \dfrac{1 + 15}{56} = .2857$

5.61   Refer to Exercise 5.60 and assume that $p = .1$ instead of $p = .5$.

a $P[x = 0] = p(0) = C_0^3 (.1)^0 (.9)^3 = .729$

$$P[x = 1] = p(1) = C_1^3 (.1)^1 (.9)^2 = .243$$

$$P[x = 2] = p(2) = C_2^3 (.1)^2 (.9)^1 = .027$$

$$P[x = 3] = p(3) = C_3^3 (.1)^3 (.9)^0 = .001$$

b   Note that the probability distribution is no longer symmetric; that is, since the probability of observing a head is so small, the probability of observing a small number of heads on three flips is increased (see the figure on the next page).

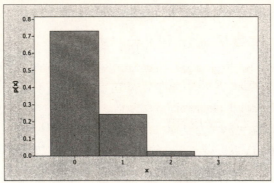

c $\mu = np = 3(.1) = .3$ and $\sigma = \sqrt{npq} = \sqrt{3(.1)(.9)} = .520$

d   The desired intervals are

$$\mu \pm \sigma = .3 \pm .520 \quad \text{or} \quad -.220 \text{ to } .820$$

$$\mu \pm 2\sigma = .3 \pm 1.04 \quad \text{or} \quad -.740 \text{ to } 1.34$$

The only value of x which falls in this first interval is $x = 0$, and the fraction of measurements in this interval will be .729. The values of $x = 0$ and $x = 1$ are enclosed by the second interval, so that $.729 + .243 = .972$ of the measurements fall within two standard deviations of the mean, consistent with both Tchebysheff's Theorem and the Empirical Rule.

5.65   Refer to Exercise 5.64. Redefine x to be the number of people who choose an interior number in the sample of $n = 20$. Then x has a binomial distribution with $p = .3$.

    a $P[x \geq 8] = 1 - P[x \leq 7] = 1 - .772 = .228$

    b   Observing eight or more people choosing an interior number is not an unlikely event, assuming that the integers are all equally likely. Therefore, there is no evidence to indicate that people are more likely to choose the interior numbers than any others.

5.69   It is given that x = number of patients with a psychosomatic problem, $n = 25$, and $p = P[\text{patient has psychosomatic problem}]$. A psychiatrist wishes to determine whether or not $p = .8$.

    a   Assuming that the psychiatrist is correct (that is, $p = .8$), the expected value of x is $E(x) = np = 25(.8) = 20$.

    b $\sigma^2 = npq = 25(.8)(.2) = 4$

    c   Given that $p = .8$, $P[x \leq 14] = .006$ from Table 1 in Appendix I.

    d   Assuming that the psychiatrist is correct, the probability of observing $x = 14$ or the more unlikely values, $x = 0, 1, 2, \ldots, 13$ is very unlikely. Hence, one of two conclusions can be drawn. Either we have observed a very unlikely event, or the psychiatrist is incorrect and p is actually less than .8. We would probably conclude that the psychiatrist is incorrect. The probability that we have made an incorrect decision is

$$P[x \le 14 \text{ given } p = .8] = .006$$

which is quite small.

5.71   Define x to be the number of students 30 years or older, with $n = 200$ and $p = P[\text{student is 30+ years}] = .25$.

    a   Since x has a binomial distribution, $\mu = np = 200(.25) = 50$ and $\sigma = \sqrt{npq} = \sqrt{200(.25)(.75)} = 6.124$.

    b   The observed value, $x = 35$, lies

$$\frac{35 - 50}{6.124} = -2.45$$

standard deviations below the mean. It is unlikely that $p = .25$.

5.75   **a**   The random variable x, the number of plants with red petals, has a binomial distribution with $n = 10$ and $p = P[\text{red petals}] = .75$.

    b   Since the value $p = .75$ is not given in Table 1, you must use the binomial formula to calculate

$$P(x \ge 9) = C_9^{10}(.75)^9(.25)^1 + C_{10}^{10}(.75)^{10}(.25)^0 = .1877 + .0563 = .2440$$

    c $P(x \le 1) = C_0^{10}(.75)^0(.25)^{10} + C_1^{10}(.75)^1(.25)^9 = .0000296$.

    d   Refer to part **c**. The probability of observing $x = 1$ or something even more unlikely $(x = 0)$ is very small – .0000296. This is a highly unlikely event if in fact $p = .75$. Perhaps there has been a nonrandom choice of seeds, or the 75% figure is not correct for this particular genetic cross.

5.79   **a**   The distribution of x is actually hypergeometric, with $N = 1200$, $n = 20$ and $M =$ number of defectives in the lot. However, since N is so large in comparison to n, the distribution of x can be closely approximated by the binomial distribution with $n = 20$ and $p = P[\text{defective}]$.

    b   If p is small, with $np < 7$, the Poisson approximation can be used.

    c   If there are 10 defectives in the lot, then $p = 10/1200 = .008333$ and $\mu = .1667$. The probability that the lot is shipped is

$$P(x = 0) \approx \frac{(.1667)^0 e^{-.1667}}{0!} = .85$$

If there are 20 defectives, $p = 20/1200$ and $\mu = .3333$. Then

$$P(x = 0) \approx \frac{(.3333)^0 e^{-.3333}}{0!} = .72$$

If there are 30 defectives, $p = 30/1200$ and $\mu = .5$. Then

$$P(x = 0) \approx \frac{(.5)^0 e^{-.5}}{0!} = .61$$

5.83   **a**   The random variable x, the number of tasters who pick the correct sample, has a binomial distribution with $n=5$ and, if there is no difference in the taste of the three samples, $p = P(\text{taster picks the correct sample}) = \dfrac{1}{3}$

b   The probability that exactly one of the five tasters chooses the latest batch as different from the others is

$$P(x = 1) = C_1^5\left(\frac{1}{3}\right)^1\left(\frac{2}{3}\right)^4 = .3292$$

c   The probability that at least one of the tasters chooses the latest batch as different from the others is

$$P(x \le 1) = 1 - P(x = 0) = 1 - C_0^5\left(\frac{1}{3}\right)^0\left(\frac{2}{3}\right)^5 = .8683$$

5.87   The random variable x has a Poisson distribution with $\mu = 2$. Use Table 2 in Appendix I or the Poisson formula to find the following probabilities.

a $P(x = 0) = \dfrac{2^0 e^{-2}}{0!} = e^{-2} = .135335$

b $P(x \le 2) = \dfrac{2^0 e^{-2}}{0!} + \dfrac{2^1 e^{-2}}{1!} + \dfrac{2^2 e^{-2}}{2!}$
$\qquad = .135335 + .270671 + .270671 = .676676$

5.91   The random variable x, the number of California homeowners with earthquake insurance, has a binomial distribution with $n = 15$ and $p = .1$.

a $P(x \ge 1) = 1 - P(x = 0) = 1 - .206 = .794$

b $P(x \ge 4) = 1 - P(x \le 3) = 1 - .944 = .056$

c   Calculate $\mu = np = 15(.1) = 1.5$ and $\sigma = \sqrt{npq} = \sqrt{15(.1)(.9)} = 1.1619$. Then approximately 95% of the values of x should lie in the interval
$$\mu \pm 2\sigma \Rightarrow 1.5 \pm 2(1.1619) \Rightarrow -.82 \text{ to } 3.82.$$
or between 0 and 3.

5.95   Use the **Calculating Binomial Probabilities** applet. The correct answers are given below.

a $P(x < 6) = 6.0(10)^{-5} = 0.00006$ **d** $P(2 < x < 6) = .5948$

b $P(x = 8) = .042$ **e** $P(x \ge 6) = 1$

c $P(x > 14) = .0207$

5.99   Define x to be the number of young adults who prefer McDonald's. Then x has a binomial distribution with $n = 100$ and $p = .5$. Use the **Calculating Binomial Probabilities** applet.

a $P(61 \le x \le 100) = .0176$

b $P(40 \le x \le 60) = .9648$

c   If 40 prefer Burger King, then 60 prefer McDonalds, and vice versa. The probability is the same as that calculated in part **b**, since $p = .5$.

6: The Normal Probability Distribution

6.3   The first few exercises are designed to provide practice for the student in evaluating areas under the normal curve. The following notes may be of some assistance.

1   Table 3, Appendix I tabulates the cumulative area under a standard normal curve to the left of a specified value of z.

2   Since the total area under the curve is one, the total area lying to the right of a specified value of z and the total area to its left must add to 1. Thus, in order to calculate a "tail area", such as the one shown in Figure 6.1, the value of $z = z_0$ will be indexed in Table 3, and the area that is obtained will be subtracted from 1. Denote the area obtained by indexing $z = z_0$ in Table 3 by $A(z_0)$ and the desired area by A. Then, in the above example, $A = 1 - A(z_0)$.

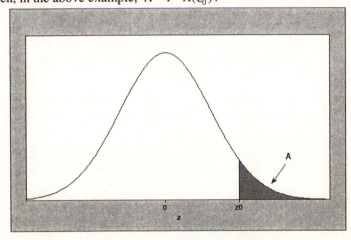

3   To find the area under the standard normal curve between two values, z_1 and z_2, calculate the difference in their cumulative areas, $A = A(z_2) - A(z_1)$.

4   Note that z, similar to x, is actually a random variable which may take on an infinite number of values, both positive and negative. Negative values of z lie to the left of the mean, $z = 0$, and positive values lie to the right.

a   It is necessary to find the area to the left of $z = 1.6$. That is, $A = A(1.6) = .9452$.

b   The area to the left of $z = 1.83$ is $A = A(1.83) = .9664$.

c $A = A(.90) = .8159$

d $A = A(4.58) \approx 1$. Notice that the values in Table 3 approach 1 as the value of z increases. When the value of z is larger than $z = 3.49$ (the largest value in the table), we can assume that the area to its left is approximately 1.

6.7   Now we are asked to find the z-value corresponding to a particular area.

a   We need to find a z_0 such that $P(z > z_0) = .025$. This is equivalent to finding an indexed area of $1 - .025 = .975$. Search the interior of Table 3 until you find the four-digit number **.9750**. The corresponding z-value is **1.96**; that is, $A(1.96) = .9750$. Therefore, $z_0 = 1.96$ is the desired z-value (see the figure below).

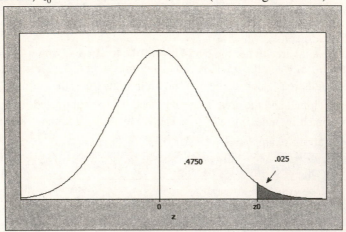

b   We need to find a z_0 such that $P(z < z_0) = .9251$ (see below). Using Table 3, we find a value such that the indexed area is .9251. The corresponding z-value is $z_0 = 1.44$.

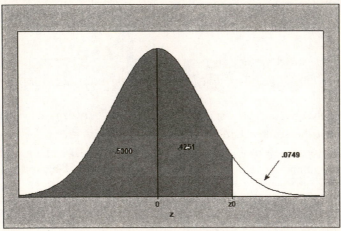

6.11   The pth percentile of the standard normal distribution is a value of z which has area $p/100$ to its left. Since all four percentiles in this exercise are greater than the 50th percentile, the value of z will all lie to the right of $z = 0$, as shown for the 90th percentile in the figure on the next page.

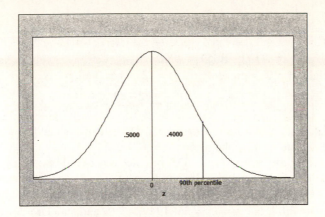

a From the figure, the area to the left of the 90th percentile is .9000. From Table 3, the appropriate value of z is closest to $z = 1.28$ with area .8997. Hence the 90th percentile is approximately $z = 1.28$.

b As in part **a**, the area to the left of the 95th percentile is .9500. From Table 3, the appropriate value of z is found using linear interpolation (see Exercise 6.9b) as $z = 1.645$. Hence the 95th percentile is $z = 1.645$.

c The area to the left of the 98th percentile is .9800. From Table 3, the appropriate value of z is closest to $z = 2.05$ with area .9798. Hence the 98th percentile is approximately $z = 2.05$.

d The area to the left of the 99th percentile is .9900. From Table 3, the appropriate value of z is closest to $z = 2.33$ with area .9901. Hence the 99th percentile is approximately $z = 2.33$.

6.15 The 99th percentile of the standard normal distribution was found in Exercise 6.11d to be $z = 2.33$. Since the relationship between the general normal random variable x and the standard normal z is $z = \dfrac{x - \mu}{\sigma}$, the corresponding percentile for this general normal random variable is found by solving for $x = \mu + z\sigma$;

$$2.33 = \frac{x - 35}{10}$$
$$x - 35 = 23.3 \quad \text{or} \quad x = 58.3$$

6.19 The random variable x, the height of a male human, has a normal distribution with $\mu = 69$ and $\sigma = 3.5$.

a A height of 6'0" represents $6(12) = 72$ inches, so that

$$P(x > 72) = P\left(z > \frac{72 - 69}{3.5}\right) = P(z > .86) = 1 - .8051 = .1949$$

b Heights of 5'8" and 6'1" represent $5(12) + 8 = 68$ and $6(12) + 1 = 73$ inches, respectively. Then

$$P(68 < x < 73) = P\left(\frac{68-69}{3.5} < z < \frac{73-69}{3.5}\right) = P(-.29 < z < 1.14)$$
$$= .8729 - .3859 = .4870$$

c A height of 6'0" represents 6(12) = 72 inches, which has a z-value of

$$z = \frac{72-69}{3.5} = .86$$

This would not be considered an unusually large value, since it is less than two standard deviations from the mean.

d The probability that a man is 6'0" or taller was found in part **a** to be .1949, which is not an unusual occurrence. However, if you define y to be the number of men in a random sample of size $n = 36$ who are 6'0" or taller, then y has a binomial distribution with mean $\mu = np = 36(.1949) = 7.02$ and standard deviation

$\sigma = \sqrt{npq} = \sqrt{36(.1949)(.8051)} = 2.38$. The value $y = 17$ lies

$$\frac{y-\mu}{\sigma} = \frac{17-7.02}{2.38} = 4.19$$

standard deviations from the mean, and would be considered an unusual occurrence for the general population of male humans. Perhaps our presidents do not represent a *random* sample from this population.

6.23   The random variable x, total weight of 8 people, has a mean of $\mu = 1200$ and a variance $\sigma^2 = 9800$. It is necessary to find $P(x > 1300)$ and $P(x > 1500)$ if the distribution of x is approximately normal. Refer to the next figure.

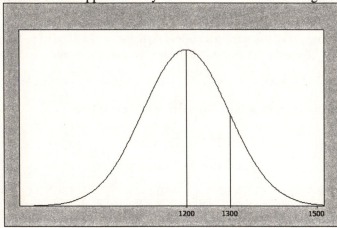

The z-value corresponding to $x_1 = 1300$ is $z_1 = \dfrac{x_1 - \mu}{\sigma} = \dfrac{1300-1200}{\sqrt{9800}} = \dfrac{100}{98.995} = 1.01$.

Hence,

$$P(x > 1300) = P(z > 1.01) = 1 - A(1.01) = 1 - .8438 = .1562.$$

Similarly, the z-value corresponding to $x_2 = 1500$ is

$$z_2 = \frac{x_2 - \mu}{\sigma} = \frac{1500 - 1200}{\sqrt{9800}} = 3.03 .$$

and $\qquad P(x > 1500) = P(z > 3.03) = 1 - A(3.03) = 1 - .9988 = .0012 .$

6.27 **a** It is given that the prime interest rate forecasts, x, are approximately normal with mean $\mu = 4.5$ and standard deviation $\sigma = 0.1$. It is necessary to determine the probability that x exceeds 4.75. Calculate

$$z = \frac{x - \mu}{\sigma} = \frac{4.75 - 4.5}{0.1} = 2.5 . \text{ Then}$$

$$P(x > 4.75) = P(z > 2.5) = 1 - .9938 = .0062 .$$

b Calculate $z = \dfrac{x - \mu}{\sigma} = \dfrac{4.375 - 4.5}{0.1} = -1.25$. Then

$$P(x < 4.375) = P(z < -1.25) = .1056 .$$

6.31 Let w be the number of words specified in the contract. Then x, the number of words in the manuscript, is normally distributed with $\mu = w + 20,000$ and $\sigma = 10,000$. The publisher would like to specify w so that

$$P(x < 100,000) = .95 .$$

As in Exercise 6.30, calculate

$$z = \frac{100,0000 - (w + 20,000)}{10,000} = \frac{80,000 - w}{10,000} .$$

Then $\qquad P(x < 100,000) = P\left(z < \dfrac{80,000 - w}{10,000} \right) = .95$. It is necessary that

$z_0 = (80,000 - w)/10,000$ be such that

$$P(z < z_0) = .95 \implies A(z_0) = .9500 \quad \text{or} \quad z_0 = 1.645 .$$

Hence,

$$\frac{80,000 - w}{10,000} = 1.645 \quad \text{or} \quad w = 63,550 .$$

6.37 **a** The normal approximation will be appropriate if both np and nq are greater than 5. For this binomial experiment,

$$np = 25(.3) = 7.5 \quad \text{and} \quad nq = 25(.7) = 17.5$$

and the normal approximation is appropriate.

b For the binomial random variable,

$$\mu = np = 7.5 \quad \text{and} \quad \sigma = \sqrt{npq} = \sqrt{25(.3)(.7)} = 2.291 .$$

c The probability of interest is the area under the binomial probability histogram corresponding to the rectangles $x = 6, 7, 8$ and 9 in the figure on the next page.

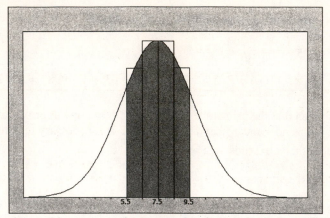

To approximate this area, use the "correction for continuity" and find the area under a normal curve with mean $\mu = 7.5$ and $\sigma = 2.291$ between $x_1 = 5.5$ and $x_2 = 9.5$. The z-values corresponding to the two values of x are

$$z_1 = \frac{5.5-7.5}{2.291} = -.87 \quad \text{and} \quad z_2 = \frac{9.5-7.5}{2.291} = .87$$

The approximating probability is

$$P(5.5 < x < 9.5) = P(-.87 < z < .87) = .8078 - .1922 = .6156 .$$

d From Table 1, Appendix I,

$$P(6 \le x \le 9) = P(x \le 9) - P(x \le 5) = .811 - .193 = .618$$

which is not too far from the approximate probability calculated in part **c**.

6.41 Using the binomial tables for $n = 20$ and $p = .3$, you can verify that

a $P(x = 5) = P(x \le 5) - P(x \le 4) = .416 - .238 = .178$

b $P(x \ge 7) = 1 - P(x \le 6) = 1 - .608 = .392$

6.45 **a** The approximating probability will be $P(x > 20.5)$ where x has a normal distribution with $\mu = 50(.32) = 16$ and $\sigma = \sqrt{50(.32)(.68)} = 3.298$. Then

$$P(x > 20.5) = P\left(z > \frac{20.5-16}{3.298}\right) = P(z > 1.36) = 1 - .9131 = .0869$$

b The approximating probability is

$$P(x < 14.5) = P\left(z < \frac{14.5-16}{3.298}\right) = P(z < -.45) = .3264$$

c If fewer than 28 students *do not* prefer cherry, then more than $50 - 28 = 22$ do prefer cherry. The approximating probability is

$$P(x > 22.5) = P\left(z > \frac{22.5-16}{3.298}\right) = P(z > 1.97) = 1 - .9756 = .0244$$

d As long as your class can be assumed to be a representative sample of all Americans, the probabilities in parts **a-c** will be accurate.

58

6.49   Define x to be the number of elections in which the taller candidate won. If Americans are not biased by height, then the random variable x has a binomial distribution with $n = 31$ and $p = .5$. Calculate

$$\mu = np = 31(.5) = 15.5 \text{ and } \sigma = \sqrt{31(.5)(.5)} = \sqrt{7.75} = 2.784$$

a   Using the normal approximation with correction for continuity, we find the area to the right of $x = 16.5$:

$$P(x > 16.5) = P\left(z > \frac{16.5 - 15.5}{2.784}\right) = P(z > .36) = 1 - .6406 = .3594$$

b   Since the occurrence of 17 out of 31 taller choices is not unusual, based on the results of part **a**, it appears that Americans do not consider height when casting a vote for a candidate.

6.53   Refer to Exercise 6.52, and let x be the number of working women who put in more than 40 hours per week on the job. Then x has a binomial distribution with $n = 50$ and $p = .62$.

a   The average value of x is $\mu = np = 50(.62) = 31$.

b   The standard deviation of x is $\sigma = \sqrt{npq} = \sqrt{50(.62)(.38)} = 3.432$.

c   The z-score for $x = 25$ is $z = \dfrac{x - \mu}{\sigma} = \dfrac{25 - 31}{3.432} = -1.75$ which is within two standard deviations of the mean. This is not considered an unusual occurrence.

6.55   **a**   The desired are A_1, as shown in the figure on the next page, is found by subtracting the cumulative areas corresponding to $z = 1.56$ and $z = 0.3$, respectively.

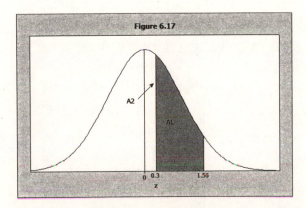

Figure 6.17

$$A_1 = A(1.56) - A(.3) = .9406 - .6179 = .3227.$$

b   The desired area is shown on the next page:
$$A_1 + A_2 = A(.2) - A(-.2) = .5793 - .4207 = .1586$$

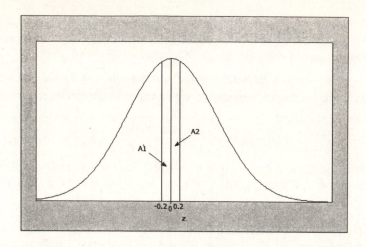

6.59 $P(-z_0 < z < z_0) = 2A(z_0) = .5000$. Hence, $A(-z_0) = \frac{1}{2}(1-.5000) = .2500$. The desired value, z_0, will be between $z_1 = .67$ and $z_2 = .68$ with associated probabilities $P_1 = .2514$ and $P_2 = .2483$. Since the desired tail area, .2500, is closer to $P_1 = .2514$, we approximate z_0 as $z_0 = .67$. The values $z = -.67$ and $z = .67$ represent the 25th and 75th percentiles of the standard normal distribution.

6.63 It is given that x is normally distributed with $\mu = 10$ and $\sigma = 3$. Let t be the guarantee time for the car. It is necessary that only 5% of the cars fail before time t (see below). That is,

$$P(x < t) = .05 \quad \text{or} \quad P\left(z < \frac{t-10}{3}\right) = .05$$

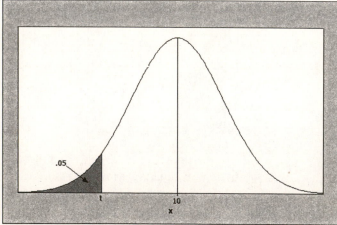

From Table 3, we know that the value of z that satisfies the above probability statement is $z = -1.645$. Hence,

$$\frac{t-10}{3} = -1.645 \quad \text{or} \quad t = 5.065 \text{ months}.$$

6.67 For this exercise $\mu = 70$ and $\sigma = 12$. The object is to determine a particular value, x_0, for the random variable x so that $P(x < x_0) = .90$ (that is, 90% of the students will finish the examination before the set time limit). Refer to the figure below.

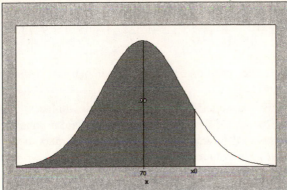

We must have

$$P(x < x_0) = P\left(z \le \frac{x_0 - 70}{12}\right) = .90$$

$$A\left(\frac{x_0 - 70}{12}\right) = .90$$

Consider $z_0 = \dfrac{x_0 - 70}{12}$. Without interpolating, the approximate value for z_0 is

$$z_0 = \frac{x_0 - 70}{12} = 1.28 \quad \text{or} \quad x_0 = 85.36$$

6.71 It is given that the random variable x (ounces of fill) is normally distributed with mean μ and standard deviation $\sigma = .3$. It is necessary to find a value of μ so that $P(x > 8) = .01$. That is, an 8-ounce cup will overflow when $x > 8$, and this should happen only 1% of the time. Then

$$P(x > 8) = P\left(z > \frac{8 - \mu}{.3}\right) = .01.$$

From Table 3, the value of z corresponding to an area (in the upper tail of the distribution) of .01 is $z_0 = 2.33$. Hence, the value of μ can be obtained by solving for μ in the following equation:

$$2.33 = \frac{8 - \mu}{.3} \quad \text{or} \quad \mu = 7.301$$

6.75 Define x = number of incoming calls that are long distance
$p = P[\text{incoming call is long distance}] = .3$

$$n = 200$$

The desired probability is $P(x \geq 50)$, where x is a binomial random variable with

$$\mu = np = 200(.3) = 60 \quad \text{and} \quad \sigma = \sqrt{npq} = \sqrt{200(.3)(.7)} = \sqrt{42} = 6.481$$

A correction for continuity is made to include the entire area under the rectangle corresponding to $x = 50$ and hence the approximation will be

$$P(x \geq 49.5) = P\left(z \geq \frac{49.5 - 60}{6.481}\right) = P(z \geq -1.62) = 1 - .0526 = .9474$$

6.79 The random variable x, the gestation time for a human baby is normally distributed with $\mu = 278$ and $\sigma = 12$.

a From Exercise 6.59, the values (rounded to two decimal places) $z = -.67$ and $z = .67$ represent the 25th and 75th percentiles of the standard normal distribution. Converting these values to their equivalents for the general random variable x using the relationship $x = \mu + z\sigma$, you have:

The lower quartile: $x = -.67(12) + 278 = 269.96$ and

The upper quartile: $x = .67(12) + 278 = 286.04$

b If you consider a month to be approximately 30 days, the value $x = 6(30) = 180$ is unusual, since it lies

$$z = \frac{x - \mu}{\sigma} = \frac{180 - 278}{12} = -8.167$$

standard deviations below the mean gestation time.

6.83 In order to implement the traditional interpretation of "curving the grades", the proportions shown in the table need to be applied to the normal curve, as shown in the figure below.

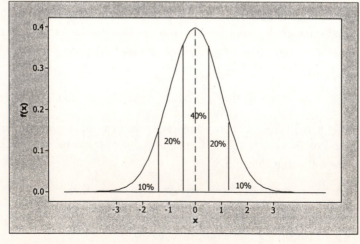

a   The C grades constitute the middle 40%, that is, 20% on either side of the mean. The lower boundary has area .3000 to its left. From Table 3, we need to find a value of z such that $A(z) = .3000$. The closest value in the table is .3015 with $z = -.52$. The upper boundary is then $z = +.52$.

b   The cutoff for the lowest D and highest B grades constitute the lower and upper boundaries of the middle 80%, that is, 40% on either side of the mean. The lower boundary has area .1000 to its left, so we need to find a value of z such that $A(z) = .1000$. The closest value in the table is .1003 with $z = -1.28$. The upper boundary is then $z = 1.28$.

6.87   Use either the **Normal Distribution Probabilities** or the **Normal Probabilities and z-scores** applets.

a $P(-2.0 < z < 2.0) = .9772 - .0228 = .9544$

b $P(-2.3 < z < -1.5) = .0668 - .0107 = .0561$

6.91   **a**   Use the **Normal Distribution Probabilities** applet. Enter 5 as the mean and 2 as the standard deviation, and the appropriate lower and upper boundaries for the probabilities you need to calculate. The probability is read from the applet as **Prob = 0.9651**.

b   Use the **Normal Probabilities and z-scores** applet. Enter 5 as the mean and 2 as the standard deviation and $x = 7.5$. Choose **One-tail** from the dropdown list and read the probability as **Prob = 0.1056**.

c   Use the **Normal Probabilities and z-scores** applet. Enter 5 as the mean and 2 as the standard deviation and $x = 0$. Choose **Cumulative** from the dropdown list and read the probability as **Prob = 0.0062**.

6.95   **a**   It is given that the scores on a national achievement test were approximately normally distributed with a mean of 540 and standard deviation of 110. It is necessary to determine how far, in standard deviations, a score of 680 departs from the mean of 540. Calculate

$$z = \frac{x - \mu}{\sigma} = \frac{680 - 540}{110} = 1.27 .$$

b   To find the percentage of people who scored higher than 680, we find the area under the standardized normal curve greater than 1.27. Using Table 3, this area is equal to

$$P(x > 680) = P(z > 1.27) = 1 - .8980 = .1020$$

Thus, approximately 10.2% of the people who took the test scored higher than 680. (The applet uses three decimal place accuracy and shows $z = 1.273$ with **Prob = 0.1016**.)

7: Sampling Distributions

7.1 You can select a simple random sample of size $n = 20$ using Table 10 in Appendix I. First choose a starting point and consider the first three digits in each number. Since the experimental units have already been numbered from 000 to 999, the first 20 can be used. The three digits OR the (three digits – 500) will identify the proper experimental unit. For example, if the three digits are 742, you should select the experimental unit numbered $742 - 500 = 242$. The probability that any three digit number is selected is $2/1000 = 1/500$. One possible selection for the sample size $n = 20$ is

242	134	173	128	399
056	412	188	255	388
469	244	332	439	101
399	156	028	238	231

7.5 If all of the town citizenry is likely to pass this corner, a sample obtained by selecting every tenth person is probably a fairly random sample.

7.9 Use a randomization scheme similar to that used in Exercise 7.1. Number each of the 50 rats from 01 to 50. To choose the 25 rats who will receive the dose of MX, select 25 two-digit random numbers from Table 10. Each two-digit number OR the (two digits – 50) will identify the proper experimental unit.

7.13 **a** The first question is more unbiased.
b Notice that the percentage favoring the new space program drops dramatically when the phrase "spending billions of dollars" is added to the question.

7.19 Regardless of the shape of the population from which we are sampling, the sampling distribution of the sample mean will have a mean μ equal to the mean of the population from which we are sampling, and a standard deviation equal to $\sigma/\sqrt{n}$.

a $\mu = 10$; $\sigma/\sqrt{n} = 3/\sqrt{36} = .5$
b $\mu = 5$; $\sigma/\sqrt{n} = 2/\sqrt{100} = .2$
c $\mu = 120$; $\sigma/\sqrt{n} = 1/\sqrt{8} = .3536$

7.22-23 For a population with $\sigma = 1$, the standard error of the mean is
$$\sigma/\sqrt{n} = 1/\sqrt{n}$$
The values of $\sigma/\sqrt{n}$ for various values of n are tabulated below and plotted on the next page. Notice that the standard error *decreases* as the sample size *increases*.

n	1	2	4	9	16	25	100
$SE(\bar{x}) = \sigma/\sqrt{n}$	1.00	.707	.500	.333	.250	.200	.100

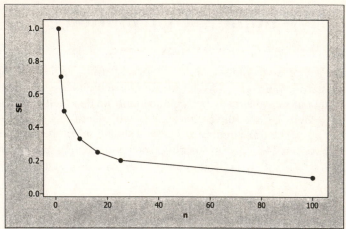

7.25   **a**   Age of equipment, technician error, technician fatigue, equipment failure, difference in chemical purity, contamination from outside sources, and so on.
b   The variability in the average measurement is measured by the standard error, $\sigma/\sqrt{n}$. In order to decrease this variability you should increase the sample size n.

7.29   **a**   The population from which we are randomly sampling $n = 35$ measurements is not necessarily normally distributed. However, the sampling distribution of $\bar{x}$ does have an approximate normal distribution, with mean μ and standard deviation $\sigma/\sqrt{n}$. The probability of interest is
$$P(|\bar{x} - \mu| < 1) = P(-1 < (\bar{x} - \mu) < 1).$$

Since $z = \dfrac{\bar{x} - \mu}{\sigma/\sqrt{n}}$ has a standard normal distribution, we need only find $\sigma/\sqrt{n}$ to approximate the above probability. Though σ is unknown, it can be approximated by $s = 12$ and $\sigma/\sqrt{n} \approx 12/\sqrt{35} = 2.028$. Then
$$P(|\bar{x} - \mu| < 1) = P(-1/2.028 < z < 1/2.028)$$
$$= P(-.49 < z < .49) = .6879 - .3121 = .3758$$

b   No. There are many possible values for x, the actual percent tax savings, as given by the probability distribution for x.

7.33   **a**   Since the original population is normally distributed, the sample mean $\bar{x}$ is also normally distributed (for any sample size) with mean μ and standard deviation
$$\sigma/\sqrt{n} \approx 0.8/\sqrt{130} = .07016$$
The z-value corresponding to $\bar{x} = 98.25$ is
$$z = \frac{\bar{x} - \mu}{\sigma/\sqrt{n}} = \frac{98.25 - 98.6}{0.8/\sqrt{130}} = -4.99$$

66

and
$$P(\bar{x} < 98.25) = P(z < -4.99) \approx 0$$

b Since the probability is extremely small, the average temperature of 98.25 degrees is very unlikely.

7.37   **a** $p = .3;\ SE(\hat{p}) = \sqrt{\dfrac{pq}{n}} = \sqrt{\dfrac{.3(.7)}{100}} = .0458$

        b $p = .1;\ SE(\hat{p}) = \sqrt{\dfrac{pq}{n}} = \sqrt{\dfrac{.1(.9)}{400}} = .015$

        c $p = .6;\ SE(\hat{p}) = \sqrt{\dfrac{pq}{n}} = \sqrt{\dfrac{.6(.4)}{250}} = .0310$

7.41   The values $SE = \sqrt{pq/n}$ for $n = 100$ and various values of p are tabulated and graphed below. Notice that SE is maximum for $p = .5$ and becomes very small for p near zero and one.

p	.01	.10	.30	.50	.70	.90	.99
$SE(\hat{p})$	.0099	.03	.0458	.05	.0458	.03	.0099

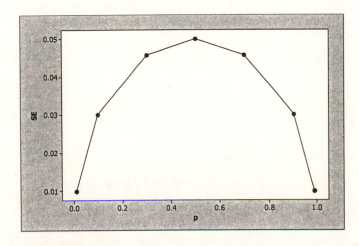

7.45   **a** The random variable $\hat{p}$, the sample proportion of brown M&Ms in a package of $n = 55$, has a binomial distribution with $n = 55$ and $p = .13$. Since $np = 7.15$ and $nq = 47.85$ are both greater than 5, this binomial distribution can be approximated by a normal distribution with mean $p = .13$ and $SE = \sqrt{\dfrac{.13(.87)}{55}} = .04535$.

        b $P(\hat{p} < .2) = P\left(z < \dfrac{.2 - .13}{.04535}\right) = P(z < 1.54) = .9382$

c $P(\hat{p} > .35) = P\left(z > \dfrac{.35 - .13}{.04535}\right) = P(z > 4.85) \approx 1 - 1 = 0$

d From the Empirical Rule (and the general properties of the normal distribution), approximately 95% of the measurements will lie within 2 (or 1.96) standard deviations of the mean:

$$p \pm 2SE \quad \Rightarrow \quad .13 \pm 2(.04535)$$

$$.13 \pm .09 \quad \text{or} \quad .04 \text{ to } .22$$

7.49 **a** The upper and lower control limits are

$$UCL = \overline{\overline{x}} + 3\dfrac{s}{\sqrt{n}} = 155.9 + 3\dfrac{4.3}{\sqrt{5}} = 155.9 + 5.77 = 161.67$$

$$LCL = \overline{\overline{x}} - 3\dfrac{s}{\sqrt{n}} = 155.9 - 3\dfrac{4.3}{\sqrt{5}} = 155.9 - 5.77 = 150.13$$

b The control chart is constructed by plotting two horizontal lines, one the upper control limit and one the lower control limit (see Figure 7.15 in the text). Values of $\overline{x}$ are plotted, and should remain within the control limits. If not, the process should be checked.

7.55 Calculate $\overline{p} = \dfrac{\sum \hat{p}_i}{k} = \dfrac{.14 + .21 + \cdots + .26}{30} = .197$. The upper and lower control limits for the p chart are then

$$UCL = \overline{p} + 3\sqrt{\dfrac{\overline{p}(1-\overline{p})}{n}} = .197 + 3\sqrt{\dfrac{.197(.803)}{100}} = .197 + .119 = .316$$

$$LCL = \overline{p} - 3\sqrt{\dfrac{\overline{p}(1-\overline{p})}{n}} = .197 - 3\sqrt{\dfrac{.197(.803)}{100}} = .197 - .119 = .078$$

7.60 **a** $C_2^4 = \dfrac{4!}{2!2!} = 6$ samples are possible.

b-c The 6 samples along with the sample means for each are shown below.

Sample	Observations	$\overline{x}$
1	6, 1	3.5
2	6, 3	4.5
3	6, 2	4.0
4	1, 3	2.0
5	1, 2	1.5
6	3, 2	2.5

d Since each of the 6 distinct values of $\overline{x}$ are equally likely (due to random sampling), the sampling distribution of $\overline{x}$ is given as

$$p(\overline{x}) = \dfrac{1}{6} \quad \text{for} \quad \overline{x} = 1.5, 2, 2.5, 3.5, 4, 4.5$$

The sampling distribution is shown below.

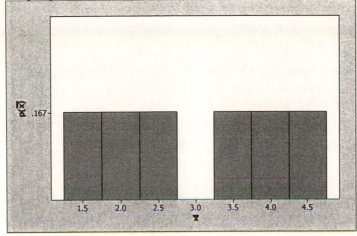

e The population mean is $\mu = (6+1+3+2)/4 = 3$. Notice that none of the samples of size $n = 2$ produce a value of $\overline{x}$ exactly equal to the population mean.

7.65 **a** To divide a group of 20 people into two groups of 10, use Table 10 in Appendix I. Assign an identification number from 01 to 20 to each person. Then select ten two digit numbers from the random number table to identify the ten people in the first group. (If the number is greater than 20, subtract multiples of 20 from the random number until you obtain a number between 01 and 20.)
b Although it is not possible to select an actual random sample from this hypothetical population, the researcher must obtain a sample that *behaves like* a random sample. A large database of some sort should be used to ensure a fairly representative sample.
c The researcher has actually selected a *convenience sample*; however, it will probably behave like a simple random sample, since a person's enthusiasm for a paid job should not affect his response to this psychological experiment.

7.71 **a** Since each cluster (a city block) is censused, this is an example of cluster sampling.
b This is a 1-in-10 systematic sample.
c The wards are the strata, and the sample is a stratified sample.
d This is a 1-in-10 systematic sample.
e This is a simple random sample from the population of all tax returns filed in the city of San Bernardino, California.

7.75 **a** The average proportion of defectives is

$$\overline{p} = \frac{.04 + .02 + \cdots + .03}{25} = .032$$

and the control limits are

$$UCL = \bar{p} + 3\sqrt{\frac{\bar{p}(1-\bar{p})}{n}} = .032 + 3\sqrt{\frac{.032(.968)}{100}} = .0848$$

and $$LCL = \bar{p} - 3\sqrt{\frac{\bar{p}(1-\bar{p})}{n}} = .032 - 3\sqrt{\frac{.032(.968)}{100}} = -.0208$$

If subsequent samples do not stay within the limits, $UCL = .0848$ and $LCL = 0$, the process should be checked.

b From part **a**, we must have $\hat{p} > .0848$.

c An erroneous conclusion will have occurred if in fact $p < .0848$ and the sample has produced $\hat{p} = .15$ by chance. One can obtain an upper bound on the probability of this particular type of error by calculating $P(\hat{p} \geq .15$ when $p = .0848)$.

7.79 Answers will vary from student to student. Paying cash for opinions will not necessarily produce a random sample of opinions of all Pepsi and Coke drinkers.

7.83 **a** The average proportion of inoperable components is
$$\bar{p} = \frac{6+7+\cdots+5}{50(15)} = \frac{75}{750} = .10$$
and the control limits are
$$UCL = \bar{p} + 3\sqrt{\frac{\bar{p}(1-\bar{p})}{n}} = .10 + 3\sqrt{\frac{.10(.90)}{50}} = .2273$$

and $$LCL = \bar{p} - 3\sqrt{\frac{\bar{p}(1-\bar{p})}{n}} = .10 - 3\sqrt{\frac{.1(.9)}{50}} = -.0272$$

If subsequent samples do not stay within the limits, $UCL = .2273$ and $LCL = 0$, the process should be checked.

7.87 **a** The theoretical mean and standard deviation of the sampling distribution of $\bar{x}$ when $n = 4$ are
$$\mu = 3.5 \quad \text{and} \quad \sigma/\sqrt{n} = 1.708/\sqrt{4} = .854$$

b-c Answers will vary from student to student. The distribution should be relatively uniform with mean and standard deviation close to those given in part **a**.

8: Large-Sample Estimation

8.1   The margin of error in estimation provides a practical upper bound to the difference between a particular estimate and the parameter which it estimates. In this chapter, the margin of error is $1.96 \times$ (standard error of the estimator).

8.5   The margin of error is $1.96\,SE = 1.96\dfrac{\sigma}{\sqrt{n}}$, where σ can be estimated by the sample standard deviation s for large values of n.

a $1.96\sqrt{\dfrac{4}{50}} = .554$ b $1.96\sqrt{\dfrac{4}{500}} = .175$ c $1.96\sqrt{\dfrac{4}{5000}} = .055$

8.9   For the estimate of p given as $\hat{p} = x/n$, the margin of error is $1.96\,SE = 1.96\sqrt{\dfrac{pq}{n}}$.

Use the estimated value given in the exercise for p.

a $1.96\sqrt{\dfrac{(.1)(.9)}{100}} = .0588$ b $1.96\sqrt{\dfrac{(.3)(.7)}{100}} = .0898$ c $1.96\sqrt{\dfrac{(.5)(.5)}{100}} = .098$

d $1.96\sqrt{\dfrac{(.7)(.3)}{100}} = .0898$ e $1.96\sqrt{\dfrac{(.9)(.1)}{100}} = .0588$

f   The largest margin of error occurs when $p = .5$.

8.13   The point estimate of μ is $\bar{x} = 39.8°$ and the margin of error with $s = 17.2$ and $n = 50$ is

$$1.96\,SE = 1.96\frac{\sigma}{\sqrt{n}} \approx 1.96\frac{s}{\sqrt{n}} = 1.96\frac{17.2}{\sqrt{50}} = 4.768$$

8.17   a   The point estimate for p is given as $\hat{p} = \dfrac{x}{n} = .78$ and the margin of error is approximately

$$1.96\sqrt{\frac{\hat{p}\hat{q}}{n}} = 1.96\sqrt{\frac{.78(.22)}{1000}} = .026$$

b   The poll's margin of error does not agree with the results of part **a**, because the sampling error was reported using the maximum margin of error using $p = .5$:

$$1.96\sqrt{\frac{\hat{p}\hat{q}}{n}} = 1.96\sqrt{\frac{.5(.5)}{1000}} = .031 \ \text{ or } \ \pm 3.1\%$$

8.21   A point estimate for the mean length of time is $\bar{x} = 19.3$, with margin of error

$$1.96\ SE = 1.96\frac{\sigma}{\sqrt{n}} \approx 1.96\frac{s}{\sqrt{n}} = 1.96\frac{5.2}{\sqrt{30}} = 1.86$$

8.23   The 90% confidence interval for μ given as

$$\bar{x} \pm 1.645\frac{\sigma}{\sqrt{n}}$$

where σ can be estimated by the sample standard deviation s for large values of n.

a $.84 \pm 1.645\sqrt{\dfrac{.086}{125}} = .84 \pm .043$ or $.797 < \mu < .883$

b $21.9 \pm 1.645\sqrt{\dfrac{3.44}{50}} = 21.9 \pm .431$ or $21.469 < \mu < 22.331$

c   Intervals constructed in this manner will enclose the true value of μ 90% of the time in repeated sampling. Hence, we are fairly confident that these particular intervals will enclose μ .

8.27   The width of a 95% confidence interval for μ is given as $1.96\dfrac{\sigma}{\sqrt{n}}$.   Hence,

a   When $n = 100$, the width is $2\left(1.96\dfrac{10}{\sqrt{100}}\right) = 2(1.96) = 3.92$.

b   When $n = 200$, the width is $2\left(1.96\dfrac{10}{\sqrt{200}}\right) = 2(1.386) = 2.772$.

c   When $n = 400$, the width is $2\left(1.96\dfrac{10}{\sqrt{400}}\right) = 2(.98) = 1.96$.

8.31   With $n = 40$, $\bar{x} = 3.7$ and $s = .5$ and $\alpha = .01$, a 99% confidence interval for μ is approximated by

$$\bar{x} \pm 2.58\frac{s}{\sqrt{n}} = 3.7 \pm 2.58\frac{.5}{\sqrt{40}} = 3.7 \pm .204 \text{ or } 3.496 < \mu < 3.904$$

In repeated sampling, 99% of all intervals constructed in this manner will enclose μ . Hence, we are fairly certain that this particular interval contains μ . (In order for this to be true, the sample must be randomly selected.)

8.35   **a**   The point estimate of p is $\hat{p} = \dfrac{x}{n} = \dfrac{68}{500} = .136$, and the approximate 95% confidence interval for p is

$$\hat{p} \pm 1.96\sqrt{\frac{\hat{p}\hat{q}}{n}} = .136 \pm 1.96\sqrt{\frac{.136(.864)}{500}} = .136 \pm .030$$

or $.106 < p < .166$.

b In order to increase the accuracy of the confidence interval, you must decrease its width. You can accomplish this by (1) increasing the sample size n, or (2) decreasing $z_{\alpha/2}$ by decreasing the confidence coefficient.

8.39 **a** When estimating the difference $\mu_1 - \mu_2$, the $(1-\alpha)100\%$ confidence interval is

$(\bar{x}_1 - \bar{x}_2) \pm z_{\alpha/2} \sqrt{\dfrac{\sigma_1^2}{n_1} + \dfrac{\sigma_2^2}{n_2}}$. Estimating σ_1^2 and σ_2^2 with s_1^2 and s_2^2, the approximate 95% confidence interval is

$(12.7 - 7.4) \pm 1.96 \sqrt{\dfrac{1.38}{35} + \dfrac{4.14}{49}} = 5.3 \pm .690 \ \ or \ \ 4.61 < \mu_1 - \mu_2 < 5.99$.

b Since the value $\mu_1 - \mu_2 = 0$ is not in the confidence interval, it is not likely that $\mu_1 = \mu_2$. You should conclude that there is a difference in the two population means.

8.43 **a** The parameter to be estimated is μ, the mean score for the posttest for all BACC classes. The 95% confidence interval is approximately

$\bar{x} \pm 1.96 \dfrac{s}{\sqrt{n}} = 18.5 \pm 1.96 \dfrac{8.03}{\sqrt{365}} = 18.5 \pm .824 \ or \ 17.676 < \mu < 19.324$

b The parameter to be estimated is μ, the mean score for the posttest for all traditional classes. The 95% confidence interval is approximately

$\bar{x} \pm 1.96 \dfrac{s}{\sqrt{n}} = 16.5 \pm 1.96 \dfrac{6.96}{\sqrt{298}} = 16.5 \pm .790 \ or \ 15.710 < \mu < 17.290$

c Now we are interested in the difference between posttest means, $\mu_1 - \mu_2$, for BACC versus traditional classes. The 95% confidence interval for $\mu_1 - \mu_2$ is approximately

$$(\bar{x}_1 - \bar{x}_2) \pm 1.96 \sqrt{\dfrac{s_1^2}{n_1} + \dfrac{s_2^2}{n_2}}$$

$$(18.5 - 16.5) \pm 1.96 \sqrt{\dfrac{8.03^2}{365} + \dfrac{6.96^2}{298}}$$

$$2.0 \pm 1.142 \ \ or \ \ .858 < (\mu_1 - \mu_2) < 3.142$$

d Since the confidence interval in part **c** has two positive endpoints, it does not contain the value $\mu_1 - \mu_2 = 0$. Hence, it is not likely that the means are equal. It appears that there is a real difference in the mean scores.

8.47 Refer to Exercise 8.18.

a The 95% confidence interval for $\mu_1 - \mu_2$ is approximately

$$(\bar{x}_1 - \bar{x}_2) \pm 1.96 \sqrt{\frac{s_1^2}{n_1} + \frac{s_2^2}{n_2}}$$

$$(170 - 160) \pm 1.96 \sqrt{\frac{17.5^2}{50} + \frac{16.5^2}{50}}$$

$$10 \pm 6.667 \quad \text{or} \quad 3.333 < (\mu_1 - \mu_2) < 16.667$$

b The 99% confidence interval for $\mu_1 - \mu_2$ is approximately

$$(\bar{x}_1 - \bar{x}_2) \pm 2.58 \sqrt{\frac{s_1^2}{n_1} + \frac{s_2^2}{n_2}}$$

$$(145 - 160) \pm 2.58 \sqrt{\frac{10^2}{50} + \frac{16.5^2}{50}}$$

$$-15 \pm 7.040 \quad \text{or} \quad -22.040 < (\mu_1 - \mu_2) < -7.960$$

c Neither of the intervals contain the value $(\mu_1 - \mu_2) = 0$. If $(\mu_1 - \mu_2) = 0$ is contained in the confidence interval, then it is not unlikely that μ_1 could equal μ_2, implying no difference in the average room rates for the two hotels. This would be of interest to the experimenter.

d Since neither confidence interval contains the value $\mu_1 - \mu_2 = 0$, it is not likely that the means are equal. You should conclude that there is a difference in the average room rates for the Marriott and Wyndham and also for the Radisson and the Wyndham chains.

8.51   **a**   Calculate $\hat{p}_1 = \frac{x_1}{n_1} = \frac{337}{800} = .42$ and $\hat{p}_2 = \frac{x_2}{n_2} = \frac{374}{640} = .58$. The approximate 90% confidence interval is

$$(\hat{p}_1 - \hat{p}_2) \pm 1.645 \sqrt{\frac{\hat{p}_1 \hat{q}_1}{n_1} + \frac{\hat{p}_2 \hat{q}_2}{n_2}}$$

$$(.42 - .58) \pm 1.645 \sqrt{\frac{.42(.58)}{800} + \frac{.58(.42)}{640}}$$

$$-.16 \pm .043 \quad \text{or} \quad -.203 < (p_1 - p_2) < -.117$$

b The two binomial samples must be random and independent and the sample sizes must be large enough that the distributions of $\hat{p}_1$ and $\hat{p}_2$ are approximately normal. Assuming that the samples are random, these conditions are met in this exercise.

8.55 **a** With $\hat{p}_1 = \dfrac{x_1}{1001} = .45$ and $\hat{p}_2 = \dfrac{x_2}{1001} = .51$. The approximate 99% confidence interval is

$$\left(\hat{p}_1 - \hat{p}_2\right) \pm 2.58 \sqrt{\frac{\hat{p}_1 \hat{q}_1}{n_1} + \frac{\hat{p}_2 \hat{q}_2}{n_2}}$$

$$\left(.45 - .51\right) \pm 2.58 \sqrt{\frac{.45(.55)}{1001} + \frac{.51(.49)}{1001}}$$

$$-.06 \pm .058 \quad \text{or} \quad -.118 < \left(p_1 - p_2\right) < -.002$$

b Since the interval in part **a** contains only negative values of $p_1 - p_2$, it is likely that $p_1 - p_2 < 0 \Rightarrow p_1 < p_2$. This would indicate that the proportion of adults who claim to be fans is higher in November than in March.

8.59 Calculate $\hat{p}_1 = \dfrac{x_1}{n_1} = \dfrac{120}{180} = .7$ and $\hat{p}_2 = \dfrac{x_2}{n_2} = \dfrac{54}{100} = .54$. The approximate 90% confidence interval is

$$\left(\hat{p}_1 - \hat{p}_2\right) \pm 1.645 \sqrt{\frac{\hat{p}_1 \hat{q}_1}{n_1} + \frac{\hat{p}_2 \hat{q}_2}{n_2}}$$

$$\left(.7 - .54\right) \pm 1.645 \sqrt{\frac{.7(.3)}{180} + \frac{.54(.46)}{100}}$$

$$.16 \pm .099 \quad \text{or} \quad .061 < \left(p_1 - p_2\right) < .259$$

Intervals constructed in this manner will enclose the true value of $p_1 - p_2$ 95% of the time in repeated sampling. Hence, we are fairly certain that this particular interval encloses $p_1 - p_2$.

8.63 Follow the instructions in the My Personal Trainer section. The answers are shown in the table below.

Type of Data	One or Two Samples	Margin of error	p or σ	Bound, B	Solve this inequality	Sample size
Binomial	One	$1.96 \sqrt{\dfrac{pq}{n}}$	$p \approx .5$	.05	$1.96 \sqrt{\dfrac{.5(.5)}{n}} \le .05$	$n \ge 385$
Quantitative	One	$1.96 \dfrac{\sigma}{\sqrt{n}}$	$\sigma \approx 10$	2	$1.96 \dfrac{10}{\sqrt{n}}$	$n \ge 97$

8.67   For the difference $\mu_1 - \mu_2$ in the population means for two quantitative populations, the 95% upper confidence bound uses $z_{.05} = 1.645$ and is calculated as

$$(\bar{x}_1 - \bar{x}_2) + 1.645\sqrt{\frac{s_1^2}{n_1} + \frac{s_2^2}{n_2}} = (12 - 10) + 1.645\sqrt{\frac{5^2}{50} + \frac{7^2}{50}}$$

$$2 + 2.00 \quad \text{or} \quad (\mu_1 - \mu_2) < 4$$

8.71   In this exercise, the parameter of interest is $p_1 - p_2$, $n_1 = n_2 = n$, and $B = .05$. Since we have no prior knowledge about p_1 and p_2, we assume the largest possible variation, which occurs if $p_1 = p_2 = .5$. Then

$$z_{\alpha/2} \times (\text{std error of } \hat{p}_1 - \hat{p}_2) \leq B$$

$$z_{.01}\sqrt{\frac{p_1 q_1}{n_1} + \frac{p_2 q_2}{n_2}} \leq .05 \quad \Rightarrow \quad 2.33\sqrt{\frac{(.5)(.5)}{n} + \frac{(.5)(.5)}{n}} \leq .05$$

$$\sqrt{n} \geq \frac{2.33\sqrt{.5}}{.05} \quad \Rightarrow \quad n \geq 1085.78 \quad \text{or} \quad n_1 = n_2 = 1086$$

8.75   The standard deviation is estimated as $s \approx R/4 = 104/4 = 26$, and .

$$2.58\sqrt{\frac{\sigma_1^2}{n_1} + \frac{\sigma_2^2}{n_2}} \leq 5 \quad \Rightarrow \quad 2.58\sqrt{\frac{26^2}{n} + \frac{26^2}{n}} \leq 5$$

$$\sqrt{n} \geq \frac{2.58\sqrt{1352}}{5} \quad \Rightarrow \quad n \geq 359.98 \quad \text{or} \quad n_1 = n_2 = 360$$

8.76   **a**   For the difference $\mu_1 - \mu_2$ in the population means this year and ten years ago, the 99% lower confidence bound uses $z_{.01} = 2.33$ and is calculated as

$$(\bar{x}_1 - \bar{x}_2) - 2.33\sqrt{\frac{s_1^2}{n_1} + \frac{s_2^2}{n_2}} = (73 - 63) - 2.33\sqrt{\frac{25^2}{400} + \frac{28^2}{400}}$$

$$10 - 4.37 \quad \text{or} \quad (\mu_1 - \mu_2) > 5.63$$

b   Since the difference in the means is positive, you can conclude that there has been a decrease in the average per-capita beef consumption over the last ten years.

8.83   **a**   The point estimate of μ is $\bar{x} = 29.1$ and the margin of error in estimation with $s = 3.9$ and $n = 64$ is

$$1.96\sigma_{\bar{x}} = 1.96\frac{\sigma}{\sqrt{n}} \approx 1.96\frac{s}{\sqrt{n}} = 1.96\left(\frac{3.9}{\sqrt{64}}\right) = .9555$$

b   The approximate 90% confidence interval is

$$\bar{x} \pm 1.645\frac{s}{\sqrt{n}} = 29.1 \pm 1.645\frac{3.9}{\sqrt{64}} = 29.1 \pm .802 \quad \text{or} \quad 28.298 < \mu < 29.902$$

Intervals constructed in this manner enclose the true value of μ 90% of the time in repeated sampling. Therefore, we are fairly certain that this particular interval encloses μ.

c The approximate 90% lower confidence bound is

$$\bar{x} - 1.28 \frac{s}{\sqrt{n}} = 29.1 - 1.28 \frac{3.9}{\sqrt{64}} = 28.48 \quad \text{or} \quad \mu > 28.48$$

d With B = .5, $\sigma \approx 3.9$, and $1 - \alpha = .95$, we must solve for n in the following inequality:

$$1.96 \frac{\sigma}{\sqrt{n}} \leq B \quad \Rightarrow \quad 1.96 \frac{3.9}{\sqrt{n}} \leq .5$$

$$\sqrt{n} \geq 15.288 \quad \Rightarrow \quad n \geq 233.723 \quad \text{or} \quad n \geq 234$$

8.87 Assuming maximum variation with $p = .5$, solve

$$1.645 \sqrt{\frac{pq}{n}} \leq .025$$

$$\sqrt{n} \geq \frac{1.645 \sqrt{.5(.5)}}{.025} = 32.9 \Rightarrow n \geq 1082.41 \quad \text{or} \quad n \geq 1083$$

8.91 **a** Define sample #1 as the sample of 482 women and sample #2 as the sample of 356 men. Then $\hat{p}_1 = .5$ and $\hat{p}_2 = .75$.

b The approximate 95% confidence interval is

$$\left(\hat{p}_1 - \hat{p}_2 \right) \pm 1.96 \sqrt{\frac{\hat{p}_1 \hat{q}_1}{n_1} + \frac{\hat{p}_2 \hat{q}_2}{n_2}}$$

$$\left(.5 - .75 \right) \pm 1.96 \sqrt{\frac{.5(.5)}{482} + \frac{.75(.25)}{356}}$$

$$-.25 \pm .063 \quad \text{or} \quad -.313 < \left(p_1 - p_2 \right) < -.187$$

c Since the value $p_1 - p_2 = 0$ is not in the confidence interval, it is unlikely that $p_1 = p_2$. You should not conclude that there is a difference in the proportion of women and men on Wall Street who have children. In fact, since all the probable values of $p_1 - p_2$ are negative, the proportion of men of Wall Street who have children appears to be larger than the proportion of women.

8.95 Assume that $\sigma = 2.5$ and the desired bound is .5. Then

$$1.96 \frac{\sigma}{\sqrt{n}} \leq B \quad \Rightarrow \quad 1.96 \frac{2.5}{\sqrt{n}} \leq .5 \quad \Rightarrow \quad n \geq 96.04 \quad \text{or} \quad n \geq 97$$

8.99 **a** If you use $p = .8$ as a conservative estimate for p, the margin of error is approximately

$$\pm 1.96\sqrt{\frac{.8(.2)}{750}} = \pm.029$$

b To reduce the margin of error in part **a** to $\pm.01$, solve for n in the equation

$$1.96\sqrt{\frac{.8(.2)}{n}} = .01 \;\Rightarrow\; \sqrt{n} = \frac{1.96(.4)}{.01} = 78.4 \;\Rightarrow n = 6146.56 \text{ or } n = 6147$$

8.103 It is assumed that $p = .2$ and that the desired bound is .01. Hence,

$$1.96\sqrt{\frac{pq}{n}} \le .01 \;\Rightarrow\; \sqrt{n} \ge \frac{1.96\sqrt{.05(.95)}}{.01} = 42.72$$

$$n \ge 1824.76 \text{ or } n \ge 1825$$

8.107 **a** The approximate 95% confidence interval for μ is

$$\bar{x} \pm 1.96\frac{s}{\sqrt{n}} = 2.962 \pm 1.96\frac{.529}{\sqrt{69}} = 2.962 \pm .125$$

or $2.837 < \mu < 3.087$.

b In order to cut the interval in half, the sample size must increase by 4. If this is done, the new half-width of the confidence interval is

$$1.96\frac{\sigma}{\sqrt{4n}} = \frac{1}{2}\left(1.96\frac{\sigma}{\sqrt{n}}\right).$$

Hence, in this case, the new sample size is $4(69) = 276$.

8.111 The approximate 98% confidence interval for μ is

$$\bar{x} \pm 2.33\frac{s}{\sqrt{n}} = 2.705 \pm 2.33\frac{.028}{\sqrt{36}} = 2.705 \pm .011$$

or $2.694 < \mu < 2.716$.

8.115 For this exercise, B $= .08$ for the binomial estimator $\hat{p}$, where $SE(\hat{p}) = \sqrt{\frac{pq}{n}}$. If $p = .3$, we have

$$1.96\sqrt{\frac{pq}{n}} \le \text{B} \Rightarrow 1.96\sqrt{\frac{.2(.8)}{n}} \le .08$$

$$\Rightarrow \sqrt{n} \ge \frac{1.96\sqrt{.2(.8)}}{.08} \Rightarrow n \ge 9.8 \text{ or } n \ge 96.04$$

or $n \ge 97$.

8.117   Use the **Interpreting Confidence Intervals** applet. Answers will vary, but the widths of all the intervals should be the same. Most of the simulations will show between 8 and 10 intervals that work correctly.

8.121   Use the **Exploring Confidence Intervals** applet.
a-b Move the slider on the right side of the applet to change the sample size. Increasing the sample size results in a smaller standard error and in a narrower interval.
c   By increasing the sample size n, you obtain more information and can obtain this more precise estimate of μ without sacrificing confidence.

9: Large-Sample Tests of Hypotheses

9.3   **a**   The critical value that separates the rejection and nonrejection regions for a right-tailed test based on a z-statistic will be a value of z (called z_α) such that $P(z > z_\alpha) = \alpha = .01$. That is, $z_{.01} = 2.33$ (see the figure below). The null hypothesis H_0 will be rejected if $z > 2.33$.

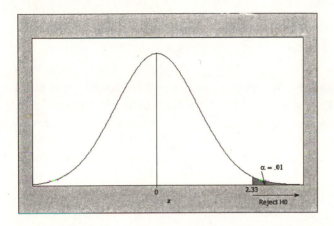

b   For a two-tailed test with $\alpha = .05$, the critical value for the rejection region cuts off $\alpha/2 = .025$ in the two tails of the z distribution in Figure 9.2, so that $z_{.025} = 1.96$. The null hypothesis H_0 will be rejected if $z > 1.96$ or $z < -1.96$ (which you can also write as $|z| > 1.96$).

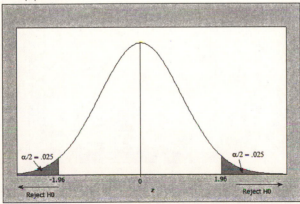

c   Similar to part **a**, with the rejection region in the lower tail of the z distribution. The null hypothesis H_0 will be rejected if $z < -2.33$.

d   Similar to part **b**, with $\alpha/2 = .005$. The null hypothesis H_0 will be rejected if $z > 2.58$ or $z < -2.58$ (which you can also write as $|z| > 2.58$).

9.6 In this exercise, the parameter of interest is μ, the population mean. The objective of the experiment is to show that the mean exceeds 2.3.

a We want to prove the alternative hypothesis that μ is, in fact, greater then 2.3. Hence, the alternative hypothesis is

$$H_a : \mu > 2.3$$

and the null hypothesis is

$$H_0 : \mu = 2.3 .$$

b The best estimator for μ is the sample average $\bar{x}$, and the test statistic is

$$z = \frac{\bar{x} - \mu_0}{\sigma / \sqrt{n}}$$

which represents the distance (measured in units of standard deviations) from $\bar{x}$ to the hypothesized mean μ. Hence, if this value is large in absolute value, one of two conclusions may be drawn. Either a very unlikely event has occurred, or the hypothesized mean is incorrect. Refer to part **a**. If $\alpha = .05$, the critical value of z that separates the rejection and non-rejection regions will be a value (denoted by z_0) such that

$$P(z > z_0) = \alpha = .05$$

That is, $z_0 = 1.645$ (see below). Hence, H_0 will be rejected if $z > 1.645$.

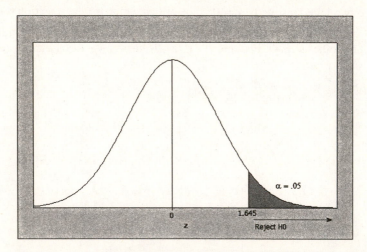

c The standard error of the mean is found using the sample standard deviation s to approximate the population standard deviation σ:

$$SE = \frac{\sigma}{\sqrt{n}} \approx \frac{s}{\sqrt{n}} = \frac{.29}{\sqrt{35}} = .049$$

d To conduct the test, calculate the value of the test statistic using the information contained in the sample. Note that the value of the true standard deviation, σ, is approximated using the sample standard deviation s.

$$z = \frac{\bar{x} - \mu_0}{\sigma/\sqrt{n}} \approx \frac{\bar{x} - \mu_0}{s/\sqrt{n}} = \frac{2.4 - 2.3}{.049} = 2.04$$

The observed value of the test statistic, $z = 2.04$, falls in the rejection region and the null hypothesis is rejected. There is sufficient evidence to indicate that $\mu > 2.3$.

9.11   **a**   In order to make sure that the average weight was one pound, you would test
$$H_0 : \mu = 1 \quad \text{versus} \quad H_a : \mu \neq 1$$
b-c The test statistic is

$$z = \frac{\bar{x} - \mu_0}{\sigma/\sqrt{n}} \approx \frac{\bar{x} - \mu_0}{s/\sqrt{n}} = \frac{1.01 - 1}{.18/\sqrt{35}} = .33$$

with p-value $= P\left(|z| > .33\right) = 2(.3707) = .7414$. Since the p-value is greater than .05, the null hypothesis should not be rejected. The manager should report that there is insufficient evidence to indicate that the mean is different from 1.

9.15   **a**   The hypothesis to be tested is
$$H_0 : \mu = 110 \quad \text{versus} \quad H_a : \mu < 110$$
and the test statistic is

$$z = \frac{\bar{x} - \mu_0}{\sigma/\sqrt{n}} \approx \frac{\bar{x} - \mu_0}{s/\sqrt{n}} = \frac{107 - 110}{13/\sqrt{100}} = -2.31$$

with p-value $= P\left(z < -2.31\right) = .0104$. To draw a conclusion from the p-value, use the guidelines for statistical significance in Section 9.3. Since the p-value is between .01 and .05, the test results are significant at the 5% level, but not at the 1% level.
b   If $\alpha = .05$, H_0 can be rejected and you can conclude that the average score improvement is less than claimed. This would be the most beneficial way for the competitor to state these conclusions.
c   If you worked for the *Princeton Review*, it would be more beneficial to conclude that there was *insufficient evidence at the 1% level* to conclude that the average score improvement is less than claimed.

9.19   The hypothesis of interest is one-tailed:
$$H_0 : \mu_1 - \mu_2 = 0 \quad \text{versus} \quad H_a : \mu_1 - \mu_2 < 0$$

The test statistic, calculated under the assumption that $\mu_1 - \mu_2 = 0$, is

$$z \approx \frac{(\bar{x}_1 - \bar{x}_2) - 0}{\sqrt{\dfrac{s_1^2}{n_1} + \dfrac{s_2^2}{n_2}}} = \frac{1.24 - 1.31}{\sqrt{\dfrac{.056}{36} + \dfrac{.054}{45}}} = -1.33$$

with the unknown σ_1^2 and σ_2^2 estimated by s_1^2 and s_2^2, respectively. The student can use one of two methods for decision making.

_p_-value approach: Calculate _p_-value $= P(z < -1.33) = .0918$. Since this _p_-value is greater than .05, the null hypothesis is not rejected. There is insufficient evidence to indicate that the mean for population 1 is smaller than the mean for population 2.
Critical value approach: The rejection region with $\alpha = .05$, is $z < -1.645$. Since the observed value of z does not fall in the rejection region, H$_0$ is not rejected. There is insufficient evidence to indicate that the mean for population 1 is smaller than the mean for population 2.

9.23   **a**   The hypothesis of interest is two-tailed:
$$H_0 : \mu_1 - \mu_2 = 0 \quad \text{versus} \quad H_a : \mu_1 - \mu_2 \neq 0$$

The test statistic, calculated under the assumption that $\mu_1 - \mu_2 = 0$, is
$$z \approx \frac{(\bar{x}_1 - \bar{x}_2) - 0}{\sqrt{\dfrac{s_1^2}{n_1} + \dfrac{s_2^2}{n_2}}} = \frac{34.1 - 36}{\sqrt{\dfrac{(5.9)^2}{100} + \dfrac{(6.0)^2}{100}}} = -2.26$$

with _p_-value $= P(|z| > 2.26) = 2(.0119) = .0238$. Since the _p_-value is less than .05, the null hypothesis is rejected. There is evidence to indicate a difference in the mean lead levels for the two sections of the city.

b   From Section 8.6, the 95% confidence interval for $\mu_1 - \mu_2$ is approximately
$$(\bar{x}_1 - \bar{x}_2) \pm 1.96 \sqrt{\dfrac{s_1^2}{n_1} + \dfrac{s_2^2}{n_2}}$$

$$(34.1 - 36) \pm 1.96 \sqrt{\dfrac{5.9^2}{100} + \dfrac{6.0^2}{100}}$$

$$-1.9 \pm 1.65 \quad \text{or} \quad -3.55 < (\mu_1 - \mu_2) < -.25$$

c   Since the value $\mu_1 - \mu_2 = 5$ or $\mu_1 - \mu_2 = -5$ is not in the confidence interval in part **b**, it is not likely that the difference will be more than 5 ppm, and hence the statistical significance of the difference is not of practical importance to the engineers.

9.27   **a**   The hypothesis of interest is two-tailed:
$$H_0 : \mu_1 - \mu_2 = 0 \quad \text{versus} \quad H_a : \mu_1 - \mu_2 \neq 0$$

and the test statistic is

$$z \approx \frac{(\bar{x}_1 - \bar{x}_2) - 0}{\sqrt{\dfrac{s_1^2}{n_1} + \dfrac{s_2^2}{n_2}}} = \frac{.94 - 2.8}{\sqrt{\dfrac{1.2^2}{36} + \dfrac{2.8^2}{26}}} = -3.18$$

with p-value $= P(|z| > 3.18) = 2(.0007) = .0014$. Since the p-value is less than .05, the null hypothesis is rejected. There is evidence to indicate a difference in the mean concentrations for these two types of sites.

b The 95% confidence interval for $\mu_1 - \mu_2$ is approximately

$$(\bar{x}_1 - \bar{x}_2) \pm 1.96 \sqrt{\frac{s_1^2}{n_1} + \frac{s_2^2}{n_2}}$$

$$(.94 - 2.8) \pm 1.96 \sqrt{\frac{1.2^2}{36} + \frac{2.8^2}{26}}$$

$$-1.86 \pm 1.15 \quad \text{or} \quad -3.01 < (\mu_1 - \mu_2) < -.71$$

Since the value $\mu_1 - \mu_2 = 0$ does not fall in the interval in part **b**, it is not likely that $\mu_1 = \mu_2$. There is evidence to indicate that the means are different, confirming the conclusion in part **a**.

9.29 **a** The hypothesis of interest is two-tailed:
$$H_0 : \mu_1 - \mu_2 = 0 \quad \text{versus} \quad H_a : \mu_1 - \mu_2 \neq 0$$

and the test statistic is

$$z \approx \frac{(\bar{x}_1 - \bar{x}_2) - 0}{\sqrt{\dfrac{s_1^2}{n_1} + \dfrac{s_2^2}{n_2}}} = \frac{98.11 - 98.39}{\sqrt{\dfrac{.7^2}{65} + \dfrac{.74^2}{65}}} = -2.22$$

with p-value $= P(|z| > 2.22) = 2(1 - .9868) = .0264$. Since the p-value is between .01 and .05, the null hypothesis is rejected, and the results are significant. There is evidence to indicate a difference in the mean temperatures for men versus women.

b Since the p-value $= .0264$, we can reject H_0 at the 5% level (p-value $< .05$), but not at the 1% level (p-value $> .01$). Using the guidelines for significance given in Section 9.3 of the text, we declare the results statistically *significant*, but not *highly significant*.

9.33 **a** The two sets of hypothesis both involve a different binomial parameter p:
$$H_0 : p = .6 \quad \text{versus} \quad H_a : p \neq .6 \text{ (part c)}$$
$$H_0 : p = .5 \quad \text{versus} \quad H_a : p < .5 \text{ (part b)}$$

b For the second test in part **a**, $x = 35$ and $n = 75$, so that $\hat{p} = \dfrac{x}{n} = \dfrac{35}{75} = .4667$, the

test statistic is

$$z = \frac{\hat{p} - p_0}{\sqrt{\frac{p_0 q_0}{n}}} = \frac{.4667 - .5}{\sqrt{\frac{.5(.5)}{75}}} = -.58$$

Since no value of α is specified in advance, we calculate
p-value $= P(z < -.58) = .2810$. Since this p-value is greater than .10, the null hypothesis is not rejected. There is insufficient evidence to contradict the claim.

c For the first test in part **a**, $x = 49$ and $n = 75$, so that $\hat{p} = \frac{x}{n} = \frac{49}{75} = .6533$, the test statistic is

$$z = \frac{\hat{p} - p_0}{\sqrt{\frac{p_0 q_0}{n}}} = \frac{.6533 - .6}{\sqrt{\frac{.6(.4)}{75}}} = .94$$

with p-value $= P(|z| > .94) = 2(.1736) = .3472$. Since the p-value is greater than .10, the null hypothesis is not rejected. There is insufficient evidence to contradict the claim.

9.37 The hypothesis of interest is
$$H_0 : p = .45 \quad \text{versus} \quad H_a : p \neq .45$$
With $\hat{p} = \frac{x}{n} = \frac{32}{80} = .4$, the test statistic is

$$z = \frac{\hat{p} - p_0}{\sqrt{\frac{p_0 q_0}{n}}} = \frac{.40 - .45}{\sqrt{\frac{.45(.55)}{80}}} = -.90$$

The rejection region is two-tailed $\alpha = .01$, or $|z| > 2.58$ and H_0 is not rejected. There is insufficient evidence to dispute the newspaper's claim.

9.41 The hypothesis of interest is
$$H_0 : p = .40 \quad \text{versus} \quad H_a : p \neq .40$$
with $\hat{p} = \frac{x}{n} = \frac{114}{300} = .38$, the test statistic is

$$z = \frac{\hat{p} - p_0}{\sqrt{\frac{p_0 q_0}{n}}} = \frac{.38 - .40}{\sqrt{\frac{.40(.60)}{300}}} = -.71$$

The rejection region with $\alpha = .05$ is $|z| > 1.96$ and the null hypothesis is not rejected. (Alternatively, we could calculate p-value $= 2P(z < -.71) = 2(.2389) = .4778$. Since this p-value is greater than .05, the null hypothesis is not rejected.) There is insufficient evidence to indicate that the proportion of households with at least one dog is different from that reported by the Humane Society.

9.45 **a** The hypothesis of interest is:
$$H_0 : p_1 - p_2 = 0 \quad \text{versus} \quad H_a : p_1 - p_2 < 0$$

Calculate $\hat{p}_1 = .36$, $\hat{p}_2 = .60$ and $\hat{p} = \dfrac{n_1 \hat{p}_1 + n_2 \hat{p}_2}{n_1 + n_2} = \dfrac{18+30}{50+50} = .48$. The test statistic is

then

$$z = \frac{\hat{p}_1 - \hat{p}_2}{\sqrt{\hat{p}\hat{q}\left(\dfrac{1}{n_1} + \dfrac{1}{n_2}\right)}} = \frac{.36 - .60}{\sqrt{.48(.52)(1/50 + 1/50)}} = -2.40$$

The rejection region, with $\alpha = .05$, is $z < -1.645$ and H_0 is rejected. There is evidence of a difference in the proportion of survivors for the two groups.
b From Section 8.7, the approximate 95% confidence interval is

$$(\hat{p}_1 - \hat{p}_2) \pm 1.96 \sqrt{\frac{\hat{p}_1 \hat{q}_1}{n_1} + \frac{\hat{p}_2 \hat{q}_2}{n_2}}$$

$$(.36 - .60) \pm 1.96 \sqrt{\frac{.36(.64)}{50} + \frac{.60(.40)}{50}}$$

$$-.24 \pm .19 \quad \text{or} \quad -.43 < (p_1 - p_2) < -.05$$

9.47 The hypothesis of interest is
$$H_0 : p_1 - p_2 = 0 \quad \text{versus} \quad H_a : p_1 - p_2 \neq 0$$

Calculate $\hat{p}_1 = \dfrac{12}{56} = .214$, $\hat{p}_2 = \dfrac{8}{32} = .25$, and $\hat{p} = \dfrac{x_1 + x_2}{n_1 + n_2} = \dfrac{12+8}{56+32} = .227$.

The test statistic is then

$$z = \frac{\hat{p}_1 - \hat{p}_2}{\sqrt{\hat{p}\hat{q}\left(\dfrac{1}{n_1} + \dfrac{1}{n_2}\right)}} = \frac{.214 - .25}{\sqrt{.227(.773)(1/56 + 1/32)}} = -.39$$

The rejection region, with $\alpha = .05$, is $|z| > 1.96$ and H_0 is not rejected. There is insufficient evidence to indicate a difference in the proportion of red M&Ms for the plain and peanut varieties. These results match the conclusions of Exercise 8.53.

9.51 The hypothesis of interest is
$$H_0 : p_1 - p_2 = 0 \quad \text{versus} \quad H_a : p_1 - p_2 > 0$$

Calculate $\hat{p}_1 = \dfrac{93}{121} = .769$, $\hat{p}_2 = \dfrac{119}{199} = .598$, and $\hat{p} = \dfrac{x_1 + x_2}{n_1 + n_2} = \dfrac{93+119}{121+199} = .6625$.

The test statistic is then

$$z = \frac{\hat{p}_1 - \hat{p}_2}{\sqrt{\hat{p}\hat{q}\left(\dfrac{1}{n_1} + \dfrac{1}{n_2}\right)}} = \frac{.769 - .598}{\sqrt{.6625(.3375)(1/121 + 1/199)}} = 3.14$$

with p-value $= P(z > 3.14) = 1 - .9992 = .0008$. Since the p-value is less than .01, the results are reported as highly significant at the 1% level of significance. There is evidence to confirm the researcher's conclusion.

9.55 The power of the test is $1 - \beta = P(\text{reject } H_0 \text{ when } H_0 \text{ is false})$. As μ gets farther from μ_0, the power of the test increases.

9.59 **a-b** Since it is necessary to prove that the average pH level is less than 7.5, the hypothesis to be tested is one-tailed:
$$H_0 : \mu = 7.5 \quad \text{versus} \quad H_a : \mu < 7.5$$
d The test statistic is
$$z = \frac{\bar{x} - \mu}{\sigma / \sqrt{n}} \approx \frac{\bar{x} - \mu}{s / \sqrt{n}} = \frac{-.2}{.2 / \sqrt{30}} = -5.477$$

and the rejection region with $\alpha = .05$ is $z < -1.645$. The observed value, $z = -5.477$, falls in the rejection region and H_0 is rejected. We conclude that the average pH level is less than 7.5.

9.63 Let p_1 be the proportion of defectives produced by machine A and p_2 be the proportion of defectives produced by machine B. The hypothesis to be tested is
$$H_0 : p_1 - p_2 = 0 \quad \text{versus} \quad H_a : p_1 - p_2 \neq 0$$
Calculate $\hat{p}_1 = \frac{16}{200} = .08$, $\hat{p}_2 = \frac{8}{200} = .04$, and $\hat{p} = \frac{x_1 + x_2}{n_1 + n_2} = \frac{16 + 8}{200 + 200} = .06$. The test statistic is then
$$z = \frac{\hat{p}_1 - \hat{p}_2}{\sqrt{\hat{p}\hat{q}\left(\frac{1}{n_1} + \frac{1}{n_2}\right)}} = \frac{.08 - .04}{\sqrt{.06(.94)(1/200 + 1/200)}} = 1.684$$

The rejection region, with $\alpha = .05$, is $|z| > 1.96$ and H_0 is not rejected. There is insufficient evidence to indicate that the machines are performing differently in terms of the percentage of defectives being produced.

9.67 The hypothesis to be tested is
$$H_0 : \mu_1 - \mu_2 = 0 \quad \text{versus} \quad H_a : \mu_1 - \mu_2 > 0$$

and the test statistic is
$$z \approx \frac{(\bar{x}_1 - \bar{x}_2) - 0}{\sqrt{\frac{s_1^2}{n_1} + \frac{s_2^2}{n_2}}} = \frac{10 - 8}{\sqrt{\frac{4.3}{40} + \frac{5.7}{40}}} = 4$$

The rejection region, with $\alpha = .05$, is one-tailed or $z > 1.645$ and the null hypothesis is rejected. There is sufficient evidence to indicate a difference in the two means. Hence, we conclude that diet I has a greater mean weight loss than diet II.

9.71   **a**   The hypothesis to be tested is
$$H_0 : \mu_1 - \mu_2 = 0 \quad \text{versus} \quad H_a : \mu_1 - \mu_2 > 0$$

and the test statistic is
$$z \approx \frac{(\bar{x}_1 - \bar{x}_2) - 0}{\sqrt{\dfrac{s_1^2}{n_1} + \dfrac{s_2^2}{n_2}}} = \frac{240 - 227}{\sqrt{\dfrac{980}{200} + \dfrac{820}{200}}} = 4.33$$

The rejection region, with $\alpha = .05$, is one-tailed or $z > 1.645$ and the null hypothesis is rejected. There is a difference in mean yield for the two types of spray.

b   An approximate 95% confidence interval for $\mu_1 - \mu_2$ is
$$(\bar{x}_1 - \bar{x}_2) \pm 1.96 \sqrt{\frac{s_1^2}{n_1} + \frac{s_2^2}{n_2}}$$

$$(240 - 227) \pm 1.96 \sqrt{\frac{980}{200} + \frac{820}{200}}$$

$$13 \pm 5.88 \quad \text{or} \quad 7.12 < (\mu_1 - \mu_2) < 18.88$$

9.75   The hypothesis to be tested is
$$H_0 : \mu = 5 \quad \text{versus} \quad H_a : \mu > 5$$
and the test statistic is
$$z = \frac{\bar{x} - \mu_0}{\sigma / \sqrt{n}} \approx \frac{\bar{x} - \mu_0}{s / \sqrt{n}} = \frac{7.2 - 5}{6.2 / \sqrt{38}} = 2.19$$

The rejection region with $\alpha = .01$ is $z > 2.33$. Since the observed value, $z = 2.19$, does not fall in the rejection region and H_0 is not rejected. The data do not provide sufficient evidence to indicate that the mean ppm of PCBs in the population of game birds exceeds the FDA's recommended limit of 5 ppm.

9.79   The hypothesis to be tested is
$$H_0 : p_1 - p_2 = 0 \quad \text{versus} \quad H_a : p_1 - p_2 \neq 0$$

Calculate $\hat{p}_1 = \dfrac{x_1}{6124} = .40$, $\hat{p}_2 = \dfrac{x_2}{5512} = .37$, and

$\hat{p} = \dfrac{x_1 + x_2}{n_1 + n_2} = \dfrac{6124(.4) + 5512(.37)}{11636} = .386$. The test statistic is then

$$z = \frac{\hat{p}_1 - \hat{p}_2}{\sqrt{\hat{p}\hat{q}\left(\dfrac{1}{n_1} + \dfrac{1}{n_2}\right)}} = \frac{.40 - .37}{\sqrt{.386(.614)(1/6124 + 1/5512)}} = 3.32$$

The rejection region for $\alpha = .01$ is $|z| > 2.58$ and the null hypothesis is rejected. There is sufficient evidence to indicate that the percentage of students who are fluent in English differs for these two districts.

9.83   **a** The parameter of interest is μ, the average daily wage of workers in a given industry. A sample of $n = 40$ workers has been drawn from a particular company within this industry and $\overline{x}$, the sample average, has been calculated. The objective is to determine whether this company pays wages different from the total industry. That is, assume that this sample of forty workers has been drawn from a hypothetical population of workers. Does this population have as an average wage $\mu = 54$, or is μ different from 54? Thus, the hypothesis to be tested is

$$H_0 : \mu = 54 \quad \text{versus} \quad H_a : \mu \neq 54.$$

b-c The test statistic is

$$z \approx \frac{\overline{x} - \mu}{s/\sqrt{n}} = \frac{51.50 - 54}{11.88/\sqrt{40}} = -1.331$$

and the **Large-Sample Test of a Population Mean** applet gives p-value $= .1832$. (Using Table 3 will produce a p-value of .1836.)

d Since $\alpha = .01$ is smaller than the p-value, .1832, H_0 cannot be rejected and we cannot conclude that the company is paying wages different from the industry average.

e Since n is greater than 30, the Central Limit Theorem will guarantee the normality of $\overline{x}$ regardless of whether the original population was normal or not.

10: Inference from Small Samples

10.1   Refer to Table 4, Appendix I, indexing *df* along the left or right margin and t_α across the top.
  a $t_{.05} = 2.015$ with 5 *df*   **b** $t_{.025} = 2.306$ with 8 *df*
  c $t_{.10} = 1.330$ with 18 *df*   **c** $t_{.025} \approx 1.96$ with 30 *df*

10.5   **a**   Using the formulas given in Chapter 2, calculate $\sum x_i = 70.5$ and $\sum x_i^2 = 499.27$. Then

$$\bar{x} = \frac{\sum x_i}{n} = \frac{70.5}{10} = 7.05$$

$$s^2 = \frac{\sum x_i^2 - \frac{\left(\sum x_i\right)^2}{n}}{n-1} = \frac{499.27 - \frac{\left(70.5\right)^2}{10}}{9} = .249444 \quad \text{and} \quad s = .4994$$

b   With $df = n-1 = 9$, the appropriate value of t is $t_{.01} = 2.821$ (from Table 4) and the 99% upper one-sided confidence bound is

$$\bar{x} + t_{.01}\frac{s}{\sqrt{n}} \quad \Rightarrow \quad 7.05 + 2.821\sqrt{\frac{.249444}{10}} \quad \Rightarrow \quad 7.05 + .446$$

or $\mu < 7.496$. Intervals constructed using this procedure will enclose μ 99% of the time in repeated sampling. Hence, we are fairly certain that this particular interval encloses μ.

c   The hypothesis to be tested is
$$H_0 : \mu = 7.5 \quad \text{versus} \quad H_a : \mu < 7.5$$
and the test statistic is

$$t = \frac{\bar{x} - \mu}{s/\sqrt{n}} = \frac{7.05 - 7.5}{\sqrt{\frac{.249444}{10}}} = -2.849$$

The rejection region with $\alpha = .01$ and $n - 1 = 9$ degrees of freedom is located in the lower tail of the *t*-distribution and is found from Table 4 as $t < -t_{.01} = -2.821$. Since the observed value of the test statistic falls in the rejection region, H_0 is rejected and we conclude that μ is less than 7.5.

d   Notice that the 99% upper one-sided confidence bound for μ does not include the value $\mu = 7.5$. This would confirm the results of the hypothesis test in part **c**, in which we concluded that μ is less than 7.5.

10.9   **a**   Similar to previous exercises. The hypothesis to be tested is
$$H_0 : \mu = 100 \quad \text{versus} \quad H_a : \mu < 100$$

Calculate $\bar{x} = \dfrac{\sum x_i}{n} = \dfrac{1797.095}{20} = 89.85475$

$$s^2 = \dfrac{\sum x_i^2 - \dfrac{\left(\sum x_i\right)^2}{n}}{n-1} = \dfrac{165{,}697.7081 - \dfrac{\left(1797.095\right)^2}{20}}{19} = 222.1150605 \quad \text{and} \quad s = 14.9035$$

The test statistic is

$$t = \dfrac{\bar{x} - \mu}{s/\sqrt{n}} = \dfrac{89.85475 - 100}{\dfrac{14.9035}{\sqrt{20}}} = -3.044$$

The critical value of t with $\alpha = .01$ and $n-1 = 19$ degrees of freedom is $t_{.01} = 2.539$ and the rejection region is $t < -2.539$. The null hypothesis is rejected and we conclude that μ is less than 100 DL.

b The 95% upper one-sided confidence bound, based on $n-1 = 19$ degrees of freedom, is

$$\bar{x} + t_{.05}\dfrac{s}{\sqrt{n}} \quad \Rightarrow \quad 89.85475 + 2.539\dfrac{14.90352511}{\sqrt{20}} \quad \Rightarrow \quad \mu < 98.316$$

This confirms the results of part **a** in which we concluded that the mean is less than 100 DL.

10.13 **a** The hypothesis to be tested is
$$H_0 : \mu = 25 \quad \text{versus} \quad H_a : \mu < 25$$

The test statistic is

$$t = \dfrac{\bar{x} - \mu_0}{s/\sqrt{n}} = \dfrac{20.3 - 25}{\dfrac{5}{\sqrt{21}}} = -4.31$$

The critical value of t with $\alpha = .05$ and $n-1 = 20$ degrees of freedom is $t_{.05} = 1.725$ and the rejection region is $t < -1.725$. Since the observed value does falls in the rejection region, H_0 is rejected, and we conclude that pre-treatment mean is less than 25.

b The 95% confidence interval based on $df = 20$ is

$$\bar{x} \pm t_{.025}\dfrac{s}{\sqrt{n}} \quad \Rightarrow \quad 26.6 \pm 2.086\dfrac{7.4}{\sqrt{21}} \quad \Rightarrow \quad 26.6 \pm 3.37$$

or $23.23 < \mu < 29.97$.

c The pre-treatment mean looks considerably smaller than the other two means.

10.17 Refer to Exercise 10.16. If we use the large sample method of Chapter 8, the large sample confidence interval is

$$\bar{x} \pm z_{.025}\dfrac{s}{\sqrt{n}} \quad \Rightarrow \quad 246.96 \pm 1.96\dfrac{46.8244}{\sqrt{50}} \quad \Rightarrow \quad 246.96 \pm 12.98$$

or $233.98 < \mu < 259.94$. The intervals are fairly similar, which is why we choose to

approximate the sampling distribution of $\dfrac{\bar{x} - \mu}{s/\sqrt{n}}$ with a z distribution when $n > 30$.

10.19 **a** $s^2 = \dfrac{(n_1 - 1)s_1^2 + (n_2 - 1)s_2^2}{n_1 + n_2 - 2} = \dfrac{9(3.4) + 3(4.9)}{10 + 4 - 2} = 3.775$

b $s^2 = \dfrac{(n_1 - 1)s_1^2 + (n_2 - 1)s_2^2}{n_1 + n_2 - 2} = \dfrac{11(18) + 20(23)}{12 + 21 - 2} = 21.2258$

10.25 **a** The hypothesis to be tested is
$$H_0 : \mu_1 - \mu_2 = 0 \quad \text{versus} \quad H_a : \mu_1 - \mu_2 \neq 0$$

From the **Minitab** printout, the following information is available:

$\bar{x}_1 = .896$ $s_1^2 = (.400)^2$ $n_1 = 14$

$\bar{x}_2 = 1.147$ $s_2^2 = (.679)^2$ $n_2 = 11$

and the test statistic is
$$t = \dfrac{(\bar{x}_1 - \bar{x}_2) - 0}{\sqrt{s^2\left(\dfrac{1}{n_1} + \dfrac{1}{n_2}\right)}} = -1.16$$

The rejection region is two-tailed, based on $n_1 + n_2 - 2 = 23$ degrees of freedom. With $\alpha = .05$, from Table 4, the rejection region is $|t| > t_{.025} = 2.069$ and H_0 is not rejected. There is not enough evidence to indicate a difference in the population means.

b It is not necessary to bound the *p*-value using Table 4, since the exact *p*-value is given on the printout as P-Value = .260.

c If you check the ratio of the two variances using the rule of thumb given in this section you will find:
$$\dfrac{\text{larger } s^2}{\text{smaller } s^2} = \dfrac{(.679)^2}{(.400)^2} = 2.88$$

which is less than three. Therefore, it is reasonable to assume that the two population variances are equal.

10.27 **a** Check the ratio of the two variances using the rule of thumb given in this section:
$$\dfrac{\text{larger } s^2}{\text{smaller } s^2} = \dfrac{2.78095}{.17143} = 16.22$$

which is greater than three. Therefore, it is not reasonable to assume that the two population variances are equal.

b You should use the unpooled *t* test with Satterthwaite's approximation to the degrees of freedom for testing
$$H_0 : \mu_1 - \mu_2 = 0 \quad \text{versus} \quad H_a : \mu_1 - \mu_2 \neq 0$$

The test statistic is

$$t = \frac{(\bar{x}_1 - \bar{x}_2) - 0}{\sqrt{\dfrac{s_1^2}{n_1} + \dfrac{s_2^2}{n_2}}} = \frac{3.73 - 4.8}{\sqrt{\dfrac{2.78095}{15} + \dfrac{.17143}{15}}} = -2.412$$

with

$$df = \frac{\left(\dfrac{s_1^2}{n_1} + \dfrac{s_2^2}{n_2}\right)^2}{\dfrac{\left(\dfrac{s_1^2}{n_1}\right)^2}{n_1 - 1} + \dfrac{\left(\dfrac{s_2^2}{n_2}\right)^2}{n_2 - 1}} = \frac{(.185397 + .0114287)^2}{.002455137 + .00000933} = 15.7$$

With $df \approx 15$, the p-value for this test is bounded between .02 and .05 so that H_0 can be rejected at the 5% level of significance. There is evidence of a difference in the mean number of uncontaminated eggplants for the two disinfectants.

10.31 **a** If swimmer 2 is faster, his(her) average time should be less than the average time for swimmer 1. Therefore, the hypothesis of interest is

$$H_0 : \mu_1 - \mu_2 = 0 \quad \text{versus} \quad H_a : \mu_1 - \mu_2 > 0$$

and the preliminary calculations are as follows:

Swimmer 1	Swimmer 2
$\sum x_{1i} = 596.46$	$\sum x_{2i} = 596.27$
$\sum x_{1i}^2 = 35576.6976$	$\sum x_{2i}^2 = 35554.1093$
$n_1 = 10$	$n_2 = 10$

Then

$$s^2 = \frac{\sum x_{1i}^2 - \dfrac{\left(\sum x_{1i}\right)^2}{n_1} + \sum x_{2i}^2 - \dfrac{\left(\sum x_{2i}\right)^2}{n_2}}{n_1 + n_2 - 2}$$

$$= \frac{35576.6976 - \dfrac{(596.46)^2}{10} + 35554.1093 - \dfrac{(596.27)^2}{10}}{5 + 5 - 2} = .03124722$$

Also, $\quad \bar{x}_1 = \dfrac{596.46}{10} = 59.646 \quad$ and $\quad \bar{x}_2 = \dfrac{596.27}{10} = 59.627$

The test statistic is

$$t = \frac{(\bar{x}_1 - \bar{x}_2) - 0}{\sqrt{s^2 \left(\dfrac{1}{n_1} + \dfrac{1}{n_2}\right)}} = \frac{59.646 - 59.627}{\sqrt{.03124722 \left(\dfrac{1}{10} + \dfrac{1}{10}\right)}} = 0.24$$

For a one-tailed test with $df = n_1 + n_2 - 2 = 18$, the p-value can be bounded using Table 4 so that p-value $> .10$, and H_0 is not rejected. There is insufficient evidence to indicate that swimmer 2's average time is still faster than the average time for swimmer 1.

10.35   **a**   The test statistic is

$$t = \frac{\bar{d} - \mu_d}{s_d/\sqrt{n}} = \frac{.3 - 0}{\sqrt{\frac{.16}{10}}} = 2.372$$

with $n - 1 = 9$ degrees of freedom. The p-value is then

$$P(|t| > 2.372) = 2P(t > 2.372) \text{ so that } P(t > 2.372) = \frac{1}{2}\text{p-value}$$

Since the value $t = 2.372$ falls between two tabled entries for $df = 9$ ($t_{.025} = 2.262$ and $t_{.01} = 2.821$), you can conclude that

$$.01 < \frac{1}{2}\text{p-value} < .025$$

$$.02 < \text{p-value} < .05$$

Since the p-value is less than $\alpha = .05$, the null hypothesis is rejected and we conclude that there is a difference in the two population means.

b   A 95% confidence interval for $\mu_1 - \mu_2 = \mu_d$ is

$$\bar{d} \pm t_{.025}\frac{s_d}{\sqrt{n}} \Rightarrow 3 \pm 2.262\sqrt{\frac{.16}{10}} \Rightarrow .3 \pm .286$$

or $.014 < (\mu_1 - \mu_2) < .586$.

c   Using $s_d^2 = .16$ and B $= .1$, the inequality to be solved is approximately

$$1.96\frac{s_d}{\sqrt{n}} \le .1$$

$$\sqrt{n} \ge \frac{1.96\sqrt{.16}}{.1} = 7.84 \Rightarrow n \ge 61.47 \text{ or } n = 62$$

Since this value of n is greater than 30, the sample size, $n = 62$ pairs, will be valid.

10.37   **a**   It is necessary to use a paired-difference test, since the two samples are not random and independent. The hypothesis of interest is

$$H_0 : \mu_1 - \mu_2 = 0 \quad \text{or} \quad H_0 : \mu_d = 0$$
$$H_a : \mu_1 - \mu_2 \ne 0 \quad \text{or} \quad H_a : \mu_d \ne 0$$

The table of differences, along with the calculation of $\bar{d}$ and s_d^2, is presented below.

d_i	.1	.1	0	.2	−.1	$\sum d_i = .3$
d_i^2	.01	.01	.00	.04	.01	$\sum d_i^2 = .07$

$$\bar{d} = \frac{\sum d_i}{n} = \frac{.3}{5} = .06 \qquad \text{and} \qquad s_d^2 = \frac{\sum d_i^2 - \dfrac{(\sum d_i)^2}{n}}{n-1} = \frac{.07 - \dfrac{(.3)^2}{5}}{4} = .013$$

The test statistic is

$$t = \frac{\bar{d} - \mu_d}{s_d/\sqrt{n}} = \frac{.06 - 0}{\sqrt{\dfrac{.013}{5}}} = 1.177$$

with $n - 1 = 4$ degrees of freedom. The rejection region with $\alpha = .05$ is $|t| > t_{.025} = 2.776$, and H$_0$ is not rejected. We cannot conclude that the means are different.

b The p-value is

$$P(|t| > 1.177) = 2P(t > 1.177) > 2(.10) = .20$$

c A 95% confidence interval for $\mu_1 - \mu_2 = \mu_d$ is

$$\bar{d} \pm t_{.025} \frac{s_d}{\sqrt{n}} \quad \Rightarrow \quad .06 \pm 2.776\sqrt{\frac{.013}{5}} \quad \Rightarrow \quad .06 \pm .142$$

or $-.082 < (\mu_1 - \mu_2) < .202$.

d In order to use the paired-difference test, it is necessary that the n paired observations be randomly selected from normally distributed populations.

10.41 **a** Each subject was presented with both signs in random order. If his reaction time in general is high, both responses will be high; if his reaction time in general is low, both responses will be low. The large variability from subject to subject will mask the variability due to the difference in sign types. The paired-difference design will eliminate the subject to subject variability.

b The hypothesis of interest is

$$H_0 : \mu_1 - \mu_2 = 0 \quad \text{or} \quad H_0 : \mu_d = 0$$
$$H_a : \mu_1 - \mu_2 \neq 0 \quad \text{or} \quad H_a : \mu_d \neq 0$$

The table of differences, along with the calculation of $\bar{d}$ and s_d^2, is presented below.

Driver	1	2	3	4	5	6	7	8	9	10	Totals
d_i	122	141	97	107	37	56	110	146	104	149	1069

$$\bar{d} = \frac{\sum d_i}{n} = \frac{1069}{10} = 106.9$$

$$s_d^2 = \frac{\sum d_i^2 - \dfrac{(\sum d_i)^2}{n}}{n-1} = \frac{126{,}561 - \dfrac{(1069)^2}{10}}{9} = 1364.98889 \quad \text{and} \quad s_d = 36.9458$$

and the test statistic is

$$t = \frac{\bar{d} - \mu_d}{s_d/\sqrt{n}} = \frac{106.9 - 0}{\dfrac{36.9458}{\sqrt{10}}} = 9.150$$

96

Since $t = 9.150$ with $df = n - 1 = 9$ is greater than the tabled value $t_{.005}$,

$$p\text{-value} < 2(.005) = .01$$

for this two tailed test and H_0 is rejected. We cannot conclude that the means are different.

c The 95% confidence interval for $\mu_1 - \mu_2 = \mu_d$ is

$$\bar{d} \pm t_{.025} \frac{s_d}{\sqrt{n}} \quad \Rightarrow \quad 106.9 \pm 2.262 \frac{36.9458}{\sqrt{10}} \quad \Rightarrow \quad 106.9 \pm 26.428$$

or $80.472 < (\mu_1 - \mu_2) < 133.328$.

10.45 **a** Use the *Minitab* printout given in the text below. The hypothesis of interest is
$$H_0 : \mu_A - \mu_B = 0 \qquad H_a : \mu_A - \mu_B > 0$$
and the test statistic is

$$t = \frac{\bar{d} - \mu_d}{s_d / \sqrt{n}} = \frac{1.4875 - 0}{\frac{1.49134}{\sqrt{8}}} = 2.82$$

The *p*-value shown in the printout is *p*-value $= .013$. Since the *p*-value is less than .05, H_0 is rejected at the 5% level of significance. We conclude that assessor A gives higher assessments than assessor B.

b A 95% lower one-sided confidence bound for $\mu_1 - \mu_2 = \mu_d$ is

$$\bar{d} - t_{.05} \frac{s_d}{\sqrt{n}} \quad \Rightarrow \quad 1.4875 - 1.895 \frac{1.49134}{\sqrt{8}} \quad \Rightarrow \quad 1.4875 - .999$$

or $(\mu_1 - \mu_2) > .4885$.

c In order to apply the paired-difference test, the 8 properties must be randomly and independently selected and the assessments must be normally distributed.

d Yes. If the individual assessments are normally distributed, then the mean of four assessments will be normally distributed. Hence, the difference $x_A - \bar{x}$ will be normally distributed and the *t* test on the differences is valid as in **c**.

10.49 For this exercise, $s^2 = .3214$ and $n = 15$. A 90% confidence interval for σ^2 will be

$$\frac{(n-1)s^2}{\chi^2_{\alpha/2}} < \sigma^2 < \frac{(n-1)s^2}{\chi^2_{(1-\alpha/2)}}$$

where $\chi^2_{\alpha/2}$ represents the value of χ^2 such that 5% of the area under the curve (shown in the figure on the next page) lies to its right. Similarly, $\chi^2_{(1-\alpha/2)}$ will be the χ^2 value such that an area .95 lies to its right.

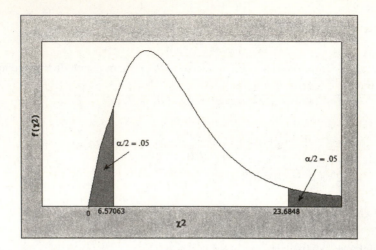

Hence, we have located one-half of α in each tail of the distribution. Indexing $\chi^2_{.05}$ and $\chi^2_{.95}$ with $n-1 = 14$ degrees of freedom in Table 5 yields

$$\chi^2_{.05} = 23.6848 \quad \text{and} \quad \chi^2_{.95} = 6.57063$$

and the confidence interval is

$$\frac{14(.3214)}{23.6848} < \sigma^2 < \frac{14(.3214)}{6.57063} \qquad \text{or} \qquad .190 < \sigma^2 < .685$$

10.53   **a**   The hypothesis to be tested is
$$H_0 : \mu = 5 \qquad H_a : \mu \neq 5$$

Calculate $\bar{x} = \dfrac{\sum x_i}{n} = \dfrac{19.96}{4} = 4.99$

$$s^2 = \frac{\sum x_i^2 - \dfrac{(\sum x_i)^2}{n}}{n-1} = \frac{99.6226 - \dfrac{(19.96)^2}{4}}{3} = .0074$$

and the test statistic is

$$t = \frac{\bar{x} - \mu_0}{s/\sqrt{n}} = \frac{4.99 - 5}{\sqrt{\dfrac{.0074}{4}}} = -.232$$

The rejection region with $\alpha = .05$ and $n-1 = 3$ degrees of freedom is found from Table 4 as $|t| > t_{.025} = 3.182$. Since the observed value of the test statistic does not fall in the rejection region, H_0 is not rejected. There is insufficient evidence to show that the mean differs from 5 mg/cc.

b   The manufacturer claims that the range of the potency measurements will equal .2. Since this range is given to equal 6σ, we know that $\sigma \approx .0333$. Then
$$H_0 : \sigma^2 = (.0333)^2 = .0011 \qquad H_a : \sigma^2 > .0011$$

The test statistic is

$$\chi^2 = \frac{(n-1)s^2}{\sigma_0^2} = \frac{3(.0074)}{.0011} = 20.18$$

and the one-tailed rejection region with $\alpha = .05$ and $n-1 = 3$ degrees of freedom is

$$\chi^2 > \chi_{.05}^2 = 7.81$$

H_0 is rejected; there is sufficient evidence to indicate that the range of the potency will exceed the manufacturer's claim.

10.57 The hypothesis of interest is $\qquad$ $H_0 : \sigma = 150 \qquad H_a : \sigma < 150$

Calculate

$$(n-1)s^2 = \sum x_i^2 - \frac{(\sum x_i)^2}{n} = 92,305,600 - \frac{(42,812)^2}{20} = 662,232.8$$

and the test statistic is $\chi^2 = \frac{(n-1)s^2}{\sigma_0^2} = \frac{662,232.8}{150^2} = 29.433$. The one-tailed rejection

region with $\alpha = .01$ and $n-1 = 19$ degrees of freedom is $\chi^2 < \chi_{.99}^2 = 7.63273$, and H_0 is not rejected. There is insufficient evidence to indicate that he is meeting his goal.

10.61 The hypothesis of interest is $\qquad$ $H_0 : \sigma_1^2 = \sigma_2^2$ versus $H_a : \sigma_1^2 \neq \sigma_2^2$

and the test statistic is

$$F = \frac{s_1^2}{s_2^2} = \frac{71^2}{69^2} = 1.059 .$$

The critical values of F for various values of α are given below using $df_1 = 15$ and $df_2 = 14$.

α	.10	.05	.025	.01	.005
F_α	2.01	2.46	2.95	3.66	4.25

Hence,

$$p\text{-value} = 2P(F > 1.059) > 2(.10) = .20$$

Since the p-value is so large, H_0 is not rejected. There is no evidence to indicate that the variances are different.

10.65 For each of the three tests, the hypothesis of interest is

$$H_0 : \sigma_1^2 = \sigma_2^2 \quad \text{versus} \quad H_a : \sigma_1^2 \neq \sigma_2^2$$

and the test statistics are

$$F = \frac{s_1^2}{s_2^2} = \frac{3.98^2}{3.92^2} = 1.03 \qquad F = \frac{s_1^2}{s_2^2} = \frac{4.95^2}{3.49^2} = 2.01 \qquad \text{and} \quad F = \frac{s_1^2}{s_2^2} = \frac{16.9^2}{4.47^2} = 14.29$$

The critical values of F for various values of α are given on the following page, using $df_1 = 9$ and $df_2 = 9$.

α	.10	.05	.025	.01	.005
F_α	2.44	3.18	4.03	5.35	6.54

Hence, for the first two tests,

$$p\text{-value} > 2(.10) = .20$$

while for the last test,

$$p\text{-value} < 2(.005) = .01$$

There is no evidence to indicate that the variances are different for the first two tests, but H_0 is rejected for the third variable. The two-sample t-test with a pooled estimate of σ^2 cannot be used for the third variable.

10.69   Paired observations are used to estimate the difference between two population means in preference to an estimation based on independent random samples selected from the two populations because of the increased information caused by blocking the observations. We expect blocking to create a large reduction in the standard deviation, if differences do exist among the blocks.

Paired observations are not always preferable. The degrees of freedom that are available for estimating σ^2 are less for paired than for unpaired observations. If there were no difference between the blocks, the paired experiment would then be less beneficial.

10.73   Since it is necessary to determine whether the injected rats drink more water than noninjected rates, the hypothesis to be tested is

$$H_0 : \mu = 22.0 \quad H_a : \mu > 22.0$$

and the test statistic is

$$t = \frac{\bar{x} - \mu_0}{s/\sqrt{n}} = \frac{31.0 - 22.0}{\dfrac{6.2}{\sqrt{17}}} = 5.985 .$$

Using the *critical value approach*, the rejection region with $\alpha = .05$ and $n - 1 = 16$ degrees of freedom is located in the upper tail of the t-distribution and is found from Table 4 as $t > t_{.05} = 1.746$. Since the observed value of the test statistic falls in the rejection region, H_0 is rejected and we conclude that the injected rats do drink more water than the noninjected rats. The 90% confidence interval is

$$\bar{x} \pm t_{.05} \frac{s}{\sqrt{n}} \quad \Rightarrow \quad 31.0 \pm 1.746 \frac{6.2}{\sqrt{17}} \quad \Rightarrow \quad 31.0 \pm 2.625$$

or $28.375 < \mu < 33.625$.

10.76   The student may use the rounded values for $\bar{x}$ and s given in the display, or he may wish to calculate $\bar{x}$ and s and use the more exact calculations for the confidence intervals. The calculations are shown on the next page.

a $\bar{x} = \dfrac{\sum x_i}{n} = \dfrac{1845}{10} = 184.5$

$$s^2 = \dfrac{\sum x_i^2 - \dfrac{(\sum x_i)^2}{n}}{n-1} = \dfrac{344{,}567 - \dfrac{(1845)^2}{10}}{9} = 462.7222$$

$s = 21.511$ and the 95% confidence interval is

$$\bar{x} \pm t_{.025}\dfrac{s}{\sqrt{n}} \;\Rightarrow\; 1.845 \pm 2.262\dfrac{21.511}{\sqrt{10}} \;\Rightarrow\; 184.5 \pm 15.4$$

or $169.1 < \mu < 199.9$.

b $\bar{x} = \dfrac{\sum x_i}{n} = \dfrac{730}{10} = 73.0$ $\qquad s^2 = \dfrac{\sum x_i^2 - \dfrac{(\sum x_i)^2}{n}}{n-1} = \dfrac{53514 - \dfrac{(730)^2}{10}}{9} = 24.8889$

$s = 4.989$ and the 95% confidence interval is

$$\bar{x} \pm t_{.025}\dfrac{s}{\sqrt{n}} \;\Rightarrow\; 73.0 \pm 2.262\dfrac{4.989}{\sqrt{10}} \;\Rightarrow\; 73.0 \pm 3.57$$

or $69.43 < \mu < 76.57$.

c $\bar{x} = \dfrac{\sum x_i}{n} = \dfrac{25.42}{10} = 2.542$

$$s^2 = \dfrac{\sum x_i^2 - \dfrac{(\sum x_i)^2}{n}}{n-1} = \dfrac{65.8398 - \dfrac{(25.42)^2}{10}}{9} = .13579556$$

$s = .3685$ and the 95% confidence interval is

$$\bar{x} \pm t_{.025}\dfrac{s}{\sqrt{n}} \;\Rightarrow\; 2.54 \pm 2.262\dfrac{.3685}{\sqrt{10}} \;\Rightarrow\; 2.54 \pm .26$$

or $2.28 < \mu < 2.80$.

d No. The relationship between the confidence intervals is not the same as the relationship between the original measurements.

10.79   Use the computing formulas or your scientific calculator to calculate

$$\bar{x} = \dfrac{\sum x_i}{n} = \dfrac{322.1}{13} = 24.777$$

$$s^2 = \dfrac{\sum x_i^2 - \dfrac{(\sum x_i)^2}{n}}{n-1} = \dfrac{8114.59 - \dfrac{(322.1)^2}{13}}{12} = 11.1619$$

$s = 3.3409$ and the 95% confidence interval is

$$\bar{x} \pm t_{.025}\frac{s}{\sqrt{n}} \Rightarrow 24.777 \pm 2.179\frac{3.3409}{\sqrt{13}} \Rightarrow 24.777 \pm 2.019$$

or $22.578 < \mu < 26.796$.

10.83 **a** The range of the first sample is 47 while the range of the second sample is only 16. There is probably a difference in the variances.

b The hypothesis of interest is

$$H_0 : \sigma_1^2 = \sigma_2^2 \quad \text{versus} \quad H_a : \sigma_1^2 \neq \sigma_2^2$$

Calculate $s_1^2 = \dfrac{177,294 - \dfrac{(838)^2}{4}}{3} = 577.6667 \qquad s_2^2 = \dfrac{192,394 - \dfrac{(1074)^2}{6}}{5} = 29.6$

and the test statistic is

$$F = \frac{s_1^2}{s_2^2} = \frac{577.6667}{29.6} = 19.516 .$$

The critical values with $df_1 = 3$ and $df_2 = 5$ are shown below from Table 6.

α	.10	.05	.025	.01	.005
F_α	3.62	5.41	7.76	12.06	16.53

Hence,

$$p\text{-value} = 2P(F > 19.516) < 2(.005) = .01$$

Since the p-value is smaller than .01, H_0 is rejected at the 1% level of significance. There is a difference in variability.

c Since the Student's t test requires the assumption of equal variance, it would be inappropriate in this instance. You should use the unpooled t test with Satterthwaite's approximation to the degrees of freedom.

10.87 A paired-difference test is used, since the two samples are not random and independent. The hypothesis of interest is

$$H_0 : \mu_1 - \mu_2 = 0 \qquad H_a : \mu_1 - \mu_2 > 0$$

and the table of differences, along with the calculation of $\bar{d}$ and s_d^2, is presented below.

Pair	1	2	3	4	Totals
d_i	−1	5	11	7	22

$$\bar{d} = \frac{\sum d_i}{n} = \frac{22}{4} = 5.5 \qquad s_d^2 = \frac{\sum d_i^2 - \dfrac{(\sum d_i)^2}{n}}{n-1} = \frac{196 - \dfrac{(22)^2}{4}}{3} = 25 \qquad \text{and}$$

$s_d = 5$

and the test statistic is

$$t = \frac{\bar{d} - \mu_d}{s_d / \sqrt{n}} = \frac{5.5 - 0}{\frac{5}{\sqrt{4}}} = 2.2$$

The one-tailed p-value with $df = 3$ can be bounded between .05 and .10. Since this value is greater than .10, H_0 is not rejected. The results are not significant; there is insufficient evidence to indicate that lack of school experience has a depressing effect on IQ scores.

10.91   The object is to determine whether or not there is a difference between the mean responses for the two different stimuli to which the people have been subjected. The samples are independently and randomly selected, and the assumptions necessary for the t test of Section 10.4 are met. The hypothesis to be tested is

$$H_0 : \mu_1 - \mu_2 = 0 \qquad H_a : \mu_1 - \mu_2 \neq 0$$

and the preliminary calculations are as follows:

$$\bar{x}_1 = \frac{15}{8} = 1.875 \quad \text{and} \quad \bar{x}_2 = \frac{21}{8} = 2.625$$

$$s_1^2 = \frac{33 - \frac{(15)^2}{8}}{7} = .69643 \quad \text{and} \quad s_2^2 = \frac{61 - \frac{(21)^2}{8}}{7} = .83929$$

Since the ratio of the variances is less than 3, you can use the pooled t test. The pooled estimator of σ^2 is calculated as

$$s^2 = \frac{(n_1 - 1)s_1^2 + (n_2 - 1)s_2^2}{n_1 + n_2 - 2} = \frac{4.875 + 5.875}{14} = .7679$$

and the test statistic is

$$t = \frac{(\bar{x}_1 - \bar{x}_2) - 0}{\sqrt{s^2 \left(\frac{1}{n_1} + \frac{1}{n_2} \right)}} = \frac{1.875 - 2.625}{\sqrt{.7679 \left(\frac{1}{8} + \frac{1}{8} \right)}} = -1.712$$

The two-tailed rejection region with $\alpha = .05$ and $df = 14$ is $|t| > t_{.025} = 2.145$, and H_0 is not rejected. There is insufficient evidence to indicate that there is a difference in means.

10.97   It is possible to test the null hypothesis $H_0 : \sigma_1^2 = \sigma_2^2$ against any one of three alternative hypotheses:

$$(1)\ \ H_a : \sigma_1^2 \neq \sigma_2^2 \qquad (2)\ \ H_a : \sigma_1^2 < \sigma_2^2 \qquad (3)\ \ H_a : \sigma_1^2 > \sigma_2^2$$

a   The first alternative would be preferred by the manager of the dairy. He does not know anything about the variability of the two machines and would wish to detect departures from equality of the type $\sigma_1^2 < \sigma_2^2$ or $\sigma_1^2 > \sigma_2^2$. These alternatives are implied in (1).

b The salesman for company A would prefer that the experimenter select the second alternative. Rejection of the null hypothesis would imply that his machine had smaller variability. Moreover, even if the null hypothesis were not rejected, there would be no evidence to indicate that the variability of the company A machine was greater than the variability of the company B machine.

c The salesman for company B would prefer the third alternative for a similar reason.

10.101 A paired-difference analysis is used. To test $H_0 : \mu_1 - \mu_2 = 0$ versus $H_a : \mu_1 - \mu_2 > 0$, where μ_2 is the mean reaction time after injection and μ_1 is the mean reaction time before injection, calculate the differences $(x_2 - x_1)$:

$$6, 1, 6, 1$$

Then $\bar{d} = \dfrac{\sum d_i}{n} = \dfrac{14}{4} = 3.5$ $s_d^2 = \dfrac{\sum d_i^2 - \dfrac{(\sum d_i)^2}{n}}{n-1} = \dfrac{74 - \dfrac{(14)^2}{4}}{3} = 8.33$ and

$s_d = 2.88675$ and the test statistic is

$$t = \frac{\bar{d} - \mu_d}{s_d / \sqrt{n}} = \frac{3.5 - 0}{\dfrac{2.88675}{\sqrt{4}}} = 2.425$$

For a one-tailed test with $df = 3$, the rejection region with $\alpha = .05$ is $t > t_{.05} = 2.353$, and H_0 is rejected. We conclude that the drug significantly increases with reaction time.

10.105 The underlying populations are ratings and can only take on the finite number of values, 1, 2, ..., 9, 10. Neither population has a normal distribution, but both are discrete. Further, the samples are not independent, since the same person is asked to rank each car design. Hence, two of the assumptions required for the Student's t test have been violated.

10.109 The *Minitab* printout below shows the summary statistics for the two samples:

Descriptive Statistics: Method 1, Method 2

Variable	N	Mean	SE Mean	StDev
Method 1	5	137.00	4.55	10.17
Method 2	5	147.20	3.29	7.36

Since the ratio of the two sample variances is less than 3, you can use the pooled t test to compare the two methods of measurement, using the remainder of the *Minitab* printout on the next page:

Two-Sample T-Test and CI: Method 1, Method 2
```
Difference = mu (Method 1) - mu (Method 2)
Estimate for difference:  -10.2000
95% CI for difference:   (-23.1506, 2.7506)
T-Test of difference = 0 (vs not =): T-Value = -1.82  P-Value = 0.107  DF = 8
Both use Pooled StDev = 8.8798
```

The test statistic is $t = -1.82$ with p-value $= .107$ and the results are not significant. There is insufficient evidence to declare a difference in the two population means.

10.113 The hypothesis to be tested is
$$H_0 : \mu = 280 \quad \text{versus} \quad H_a : \mu > 280$$
The test statistic is
$$t = \frac{\bar{x} - \mu}{s/\sqrt{n}} = \frac{358 - 280}{\dfrac{54}{\sqrt{10}}} = 4.57$$

The critical value of t with $\alpha = .01$ and $n - 1 = 9$ degrees of freedom is $t_{.01} = 2.821$ and the rejection region is $t > 2.821$. Since the observed value falls in the rejection region, H_0 is rejected. There is sufficient evidence to indicate that the average number of calories is greater than advertised.

10.119 Use the **Interpreting Confidence Intervals** applet. Answers will vary from student to student. The widths of the ten intervals will not be the same, since the value of s changes with each new sample. The student should find that approximately 95% of the intervals in the first applet contain μ, while roughly 99% of the intervals in the second applet contain μ.

10.123 Use the **Two Sample t Test: Independent Samples** applet. The hypothesis to be tested concerns the differences between mean recovery rates for the two surgical procedures. Let μ_1 be the population mean for Procedure I and μ_2 be the population mean for Procedure II. The hypothesis to be tested is
$$H_0 : \mu_1 - \mu_2 = 0 \quad \text{versus} \quad H_a : \mu_1 - \mu_2 \neq 0$$

Since the ratio of the variances is less than 3, you can use the pooled t test. Enter the appropriate statistics into the applet and you will find that test statistic is
$$t = \frac{(\bar{x}_1 - \bar{x}_2) - 0}{\sqrt{s^2 \left(\dfrac{1}{n_1} + \dfrac{1}{n_2} \right)}} = -3.33$$

with a two-tailed p-value of .0030. Since the p-value is very small, H_0 can be rejected for any value of α greater than .003 and the results are judged highly significant. There is sufficient evidence to indicate a difference in the mean recovery rates for the two procedures.

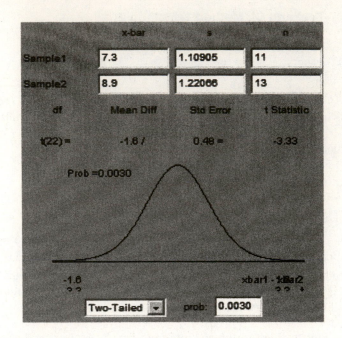

	x-bar	s	n
Sample1	7.3	1.10905	11
Sample2	8.9	1.22066	13

df	Mean Diff	Std Error	t Statistic
t(22) =	-1.6 /	0.48 =	-3.33

Prob =0.0030

-1.6

xbar1 - xbar2

Two-Tailed   prob: 0.0030

11: The Analysis of Variance

11.1   In comparing 6 populations, there are $k-1$ degrees of freedom for treatments and $n = 6(10) = 60$. The ANOVA table is shown below.

Source	df
Treatments	5
Error	54
Total	59

11.4   Similar to Exercise 11.1. With $n = 4(6) = 24$ and $k = 4$, the sources of variation and associated *df* are shown below.

Source	df
Treatments	3
Error	20
Total	23

11.7   The following preliminary calculations are necessary:

$$T_1 = 14 \quad T_2 = 19 \quad T_3 = 5 \quad G = 38$$

a $CM = \dfrac{\left(\sum x_{ij}\right)^2}{n} = \dfrac{(38)^2}{14} = 103.142857$

Total $SS = \sum x_{ij}^2 - CM = 3^2 + 2^2 + \cdots + 2^2 + 1^2 - CM = 130 - 103.142857 = 26.8571$

b $SST = \sum \dfrac{T_i^2}{n_i} - CM = \dfrac{14^2}{5} + \dfrac{19^2}{5} + \dfrac{5^2}{4} - CM = 117.65 - 103.142857 = 14.5071$

and $MST = \dfrac{SST}{k-1} = \dfrac{14.5071}{2} = 7.2536$

c   By subtraction, $SSE = \text{Total } SS - SST = 26.8571 - 14.5071 = 12.3500$ and the degrees of freedom, by subtraction, are $13 - 2 = 11$. Then

$$MSE = \dfrac{SSE}{11} = \dfrac{12.3500}{11} = 1.1227$$

d   The information obtained in parts **a-c** is consolidated in an ANOVA table.

Source	df	SS	MS
Treatments	2	14.5071	7.2536
Error	11	12.3500	1.1227
Total	13	26.8571	

e   The hypothesis to be tested is

$H_0 : \mu_1 = \mu_2 = \mu_3$ versus H_a : at least one pair of means are different

f The rejection region for the test statistic $F = \dfrac{\text{MST}}{\text{MSE}} = \dfrac{7.2536}{1.1227} = 6.46$ is based on an

F-distribution with 2 and 11 degrees of freedom. The critical values of F for bounding the p-value for this one-tailed test are shown below.

α	.10	.05	.025	.01	.005
F_α	2.86	3.98	5.26	7.21	8.91

Since the observed value $F = 6.46$ is between $F_{.01}$ and $F_{.025}$,

$$.01 < p\text{-value} < .025$$

and H_0 is rejected at the 5% level of significance. There is a difference among the means.

11.10 **a** The following preliminary calculations are necessary:
$$T_1 = 380 \quad T_2 = 199 \quad T_3 = 261 \quad G = 840$$

$$CM = \frac{\left(\sum x_{ij}\right)^2}{n} = \frac{(840)^2}{11} = 64{,}145.4545$$

Total $SS = \sum x_{ij}^2 - CM = 65{,}286 - CM = 1140.5455$

$$SST = \sum \frac{T_i^2}{n_i} - CM = \frac{380^2}{5} + \frac{199^2}{3} + \frac{261^2}{3} - CM = 641.87883$$

Calculate $MS = SS/df$ and consolidate the information in an ANOVA table.

Source	df	SS	MS
Treatments	2	641.8788	320.939
Error	8	498.6667	62.333
Total	10	1140.5455	

b The hypothesis to be tested is

$H_0 : \mu_1 = \mu_2 = \mu_3$ versus H_a : at least one pair of means are different
and the F test to detect a difference in mean student response is

$$F = \frac{\text{MST}}{\text{MSE}} = 5.15 .$$

The rejection region with $\alpha = .05$ and 2 and 8 df is $F > 4.46$ and H_0 is rejected. There is a significant difference in mean response due to the three different methods.

11.13 **a** We would be reasonably confident that the data satisfied the normality assumption because each measurement represents the average of 10 continuous measurements. The Central Limit Theorem assures us that this mean will be approximately normally distributed.
b We have a completely randomized design with four treatments, each containing 6 measurements. The analysis of variance table is given in the **Minitab** printout. The F test is

$$F = \frac{\text{MST}}{\text{MSE}} = \frac{6.580}{.115} = 57.38$$

with p-value $= .000$ (in the column marked "P"). Since the p-value is very small (less than .01), H_0 is rejected. There is a significant difference in the mean leaf length among the four locations with $P < .01$ or even $P < .001$.

c The hypothesis to be tested is $H_0 : \mu_1 = \mu_4$ versus $H_a : \mu_1 \neq \mu_4$ and the test statistic is

$$t = \frac{\bar{x}_1 - \bar{x}_4}{\sqrt{MSE\left(\frac{1}{n_1} + \frac{1}{n_4}\right)}} = \frac{6.0167 - 3.65}{\sqrt{.115\left(\frac{1}{6} + \frac{1}{6}\right)}} = 12.09$$

The p-value with $df = 20$ is $2P(t > 12.09)$ is bounded (using Table 4) as

$$p\text{-value} < 2(.005) = .01$$

and the null hypothesis is reject. We conclude that there is a difference between the means.

d The 99% confidence interval for $\mu_1 - \mu_4$ is

$$(\bar{x}_1 - \bar{x}_4) \pm t_{.005}\sqrt{MSE\left(\frac{1}{n_1} + \frac{1}{n_4}\right)}$$

$$(6.0167 - 3.65) \pm 2.845\sqrt{.115\left(\frac{1}{6} + \frac{1}{6}\right)}$$

$$2.367 \pm .557 \quad \text{or} \quad 1.810 < \mu_1 - \mu_4 < 2.924$$

e When conducting the t tests, remember that the stated confidence coefficients are based on random sampling. If you looked at the data and only compared the largest and smallest sample means, the randomness assumption would be disturbed.

11.17 **a** The design is a completely randomized design (four independent samples).
b The following preliminary calculations are necessary:

$$T_1 = 1211 \quad T_2 = 1074 \quad T_3 = 1158 \quad T_4 = 1243 \quad G = 4686$$

$$CM = \frac{\left(\sum x_{ij}\right)^2}{n} = \frac{(4686)^2}{20} = 1,097,929.8$$

Total $SS = \sum x_{ij}^2 - CM = 1,101,862 - CM = 3932.2$

$$SST = \sum \frac{T_i^2}{n_i} - CM = \frac{1211^2}{5} + \frac{1074^2}{5} + \frac{1158^2}{5} + \frac{1243^2}{5} - CM = 3272.2$$

Calculate $MS = SS/df$ and consolidate the information in an ANOVA table.

Source	df	SS	MS
Treatments	3	3272.2	1090.7333
Error	16	660	41.25
Total	19	3932.2	

c The hypothesis to be tested is
$$H_0 : \mu_1 = \mu_2 = \mu_3 = \mu_4 \quad \text{versus} \quad H_a : \text{at least one pair of means are different}$$

and the F test to detect a difference in average prices is

$$F = \frac{MST}{MSE} = 26.44 .$$

The rejection region with $\alpha = .05$ and 3 and 16 df is approximately $F > 3.24$ and H_0 is rejected. There is enough evidence to indicate a difference in the average prices for the four states.

11.21   **a** $\omega = q_{.05}(4,12)\dfrac{s}{\sqrt{5}} = 4.20\dfrac{s}{\sqrt{5}} = 1.878s$

         b $\omega = q_{.01}(6,12)\dfrac{s}{\sqrt{8}} = 6.10\dfrac{s}{\sqrt{8}} = 2.1567s$

11.25   The design is completely randomized with 3 treatments and 5 replications per treatment. The *Minitab* printout below shows the analysis of variance for this experiment.

One-way ANOVA: mg/dl versus Lab

```
Source   DF      SS      MS      F       P
Lab       2     42.6    21.3   0.60   0.562
Error    12    422.5    35.2
Total    14    465.0

S = 5.933    R-Sq = 9.15%    R-Sq(adj) = 0.00%

                              Individual 95% CIs For Mean Based on
                              Pooled StDev
Level   N    Mean   StDev    --+---------+---------+---------+-------
1       5  108.86    7.47                   (-------------*--------------)
2       5  105.04    6.01    (--------------*-------------)
3       5  105.60    3.70      (-------------*-------------)
                              --+---------+---------+---------+-------
                            100.0      104.0     108.0     112.0
Pooled StDev = 5.93

Tukey 95% Simultaneous Confidence Intervals
All Pairwise Comparisons among Levels of Lab
Individual confidence level = 97.94%

Lab = 1 subtracted from:
Lab    Lower   Center  Upper    +---------+---------+---------+-------
2    -13.824   -3.820  6.184    (--------------*-------------)
3    -13.264   -3.260  6.744     (-------------*--------------)
                                 +---------+---------+---------+-------
                               -14.0      -7.0      0.0       7.0

Lab = 2 subtracted from:
Lab    Lower   Center  Upper    +---------+---------+---------+-------
3     -9.444    0.560 10.564                (-------------*-------------)
                                 +---------+---------+---------+-------
                               -14.0      -7.0      0.0       7.0
```

a   The analysis of variance F test for $H_0 : \mu_1 = \mu_2 = \mu_3$ is $F = .60$ with p-value $= .562$. The results are not significant and H_0 is not rejected. There is insufficient evidence to indicate a difference in the treatment means.

b   Since the treatment means are not significantly different, there is no need to use Tukey's test to search for the pairwise differences. Notice that all three intervals generated by ***Minitab*** contain zero, indicating that the pairs cannot be judged different.

11.29   Refer to Exercise 11.28. The given sums of squares are inserted and missing entries found by subtraction. The mean squares are found as $MS = SS/df$.

Source	df	SS	MS	F
Treatments	2	11.4	5.70	4.01
Blocks	5	17.1	3.42	2.41
Error	10	14.2	1.42	
Total	17	42.7		

11.33   Use ***Minitab*** to obtain an ANOVA printout, or use the following calculations:

$$CM = \frac{\left(\sum x_{ij}\right)^2}{n} = \frac{(113)^2}{12} = 1064.08333$$

$$\text{Total SS} = \sum x_{ij}^2 - CM = 6^2 + 10^2 + \cdots + 14^2 - CM = 1213 - CM = 148.91667$$

$$SST = \sum \frac{T_j^2}{3} - CM = \frac{22^2 + 34^2 + 27^2 + 30^2}{3} - CM = 25.58333$$

$$SSB = \sum \frac{B_i^2}{4} - CM = \frac{33^2 + 25^2 + 55^2}{4} - CM = 120.66667 \text{ and}$$

$$SSE = \text{Total SS} - SST - SSB = 2.6667$$

Calculate $MS = SS/df$ and consolidate the information in an ANOVA table.

Source	df	SS	MS	F
Treatments	3	25.5833	8.5278	19.19
Blocks	2	120.6667	60.3333	135.75
Error	6	2.6667	0.4444	
Total	11	148.9167		

a   To test the difference among treatment means, the test statistic is

$$F = \frac{MST}{MSE} = \frac{8.528}{.4444} = 19.19$$

and the rejection region with $\alpha = .05$ and 3 and 6 df is $F > 4.76$. There is a significant difference among the treatment means.

b   To test the difference among block means, the test statistic is

$$F = \frac{MSB}{MSE} = \frac{60.3333}{.4444} = 135.75$$

111

and the rejection region with $\alpha = .05$ and 2 and 6 df is $F > 5.14$. There is a significant difference among the block means.

c With $k = 4$, $df = 6$, $n_t = 3$,

$$\omega = q_{.01}(4,6)\frac{\sqrt{\text{MSE}}}{\sqrt{n_t}} = 7.03\sqrt{\frac{.4444}{3}} = 2.71$$

The ranked means are shown below.

7.33	9.00	10.00	11.33
$\overline{x}_1$	$\overline{x}_3$	$\overline{x}_4$	$\overline{x}_2$

d The 95% confidence interval is

$$\left(\overline{x}_A - \overline{x}_B\right) \pm t_{.025}\sqrt{\text{MSE}\left(\frac{2}{b}\right)}$$

$$(7.333 - 11.333) \pm 2.447\sqrt{.4444\left(\frac{2}{3}\right)}$$

$$-4 \pm 1.332 \quad \text{or} \quad -5.332 < \mu_A - \mu_B < -2.668$$

e Since there is a significant difference among the block means, blocking has been effective. The variation due to block differences can be isolated using the randomized block design.

11.37 Similar to previous exercises. The **Minitab** printout for this randomized block experiment is shown below.

Two-way ANOVA: Measurements versus Blocks, Chemicals

```
Source       DF      SS       MS       F       P
Blocks        2   7.1717   3.58583   40.21   0.000
Chemicals     3   5.2000   1.73333   19.44   0.002
Error         6   0.5350   0.08917
Total        11  12.9067
S = 0.2986    R-Sq = 95.85%    R-Sq(adj) = 92.40%


                    Individual 95% CIs For Mean Based on Pooled
                    StDev
Blocks    Mean      +---------+---------+---------+---------
1        10.875     (----*-----)
2        12.700                                  (----*-----)
3        12.225                         (-----*----)
                    +---------+---------+---------+---------
                    10.50     11.20     11.90     12.60


                    Individual 95% CIs For Mean Based on
                    Pooled StDev
Chemicals   Mean    ------+---------+---------+---------+---
1         11.4000   (-----*-----)
2         12.3333                   (-----*-----)
3         11.2000 (-----*-----)
4         12.8000                            (-----*-----)
                    ------+---------+---------+---------+---
                        11.20     11.90     12.60     13.30
```

Both the treatment and block means are significantly different. Since the four chemicals represent the treatments in this experiment, Tukey's test can be used to determine where the differences lie:

$$\omega = q_{.05}(4,6)\frac{\sqrt{MSE}}{\sqrt{n_t}} = 4.90\sqrt{\frac{.08917}{3}} = .845$$

The ranked means are shown below.

11.20	11.40	12.33	12.80
$\overline{x}_3$	$\overline{x}_1$	$\overline{x}_2$	$\overline{x}_4$

The chemical falls into two significantly different groups – A and C versus B and D.

11.41 A randomized block design has been used with "estimators" as treatments and "construction job" as the block factor. The analysis of variance table is found in the *Minitab* printout below.

Two-way ANOVA: Cost versus Estimator, Job

```
Source       DF        SS        MS        F       P
Estimator     2    10.8617    5.4308    7.20    0.025
Job           3    37.6073   12.5358   16.61    0.003
Error         6     4.5283    0.7547
Total        11    52.9973
S = 0.8687    R-Sq = 91.46%    R-Sq(adj) = 84.34%
```

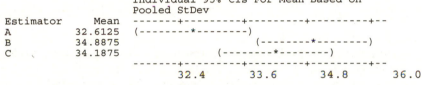

```
                    Individual 95% CIs For Mean Based on
                    Pooled StDev
Estimator    Mean    -------+---------+---------+---------+--
A          32.6125   (--------*--------)
B          34.8875                      (--------*--------)
C          34.1875               (--------*--------)
                    -------+---------+---------+---------+--
                       32.4      33.6      34.8      36.0
```

Both treatments and blocks are significant. The treatment means can be further compared using Tukey's test with

$$\omega = q_{.05}(3,6)\frac{\sqrt{MSE}}{\sqrt{n_t}} = 4.34\sqrt{\frac{.7547}{4}} = 1.885$$

The ranked means are shown below.

32.6125	34.1875	34.8875
$\overline{x}_A$	$\overline{x}_C$	$\overline{x}_B$

Estimators A and B show a significant difference in average costs.

11.45 **a-b** There are $4\times5 = 20$ treatments and $4\times5\times3 = 60$ total observations.
c In a factorial experiment, variation due to the interaction $A\times B$ is isolated from SSE. The sources of variation and associated degrees of freedom are given on the next page.

Source	df
A	3
B	4
A×B	12
Error	40
Total	59

11.49 **a** Based on the fact that the mean response for the two levels of factor B behaves very differently depending on the level of factor A under investigation, there is a strong interaction present between factors A and B.

b The test statistic for interaction is

$F = \text{MS}(\text{AB})/\text{MSE} = 37.85$ with p-value $= .000$ from the **Minitab** printout. There is evidence of a significant interaction. That is, the effect of factor A depends upon the lvel of factor B at which A is measured.

c In light of this type of interaction, the main effect means (averaged over the levels of the other factor) differ only slightly. Hence, a test of the main-effect terms produces a non-significant result.

d No. A significant interaction indicates that the effect of one factor depends upon the level of the other. Each factor-level combination should be investigated individually.

e Answers will vary.

11.53 **a** The design is a 2×4 factorial experiment with $r = 5$ replications. There are two factors, Gender and School, one at two levels and one at four levels.

b The analysis of variance table can be found using a computer printout or the following calculations:

Gender	Schools 1	2	3	4	Total
Male	2919	3257	3330	2461	11967
Female	3082	3629	3344	2410	12465
Total	6001	6886	6674	4871	24432

$$\text{CM} = \frac{24432^2}{40} = 14923065.6$$

$$\text{Total SS} = 15281392 - \text{CM} = 358326.4$$

$$\text{SSG} = \frac{11967^2 + 12465^2}{20} - \text{CM} = 6200.1$$

$$\text{SS}(\text{Sc}) = \frac{6001^2 + 6886^2 + 6674^2 + 4871^2}{10} - \text{CM} = 246725.8$$

$$\text{SS}(\text{G} \times \text{Sc}) = \frac{2919^2 + 3257^2 + \cdots + 2410^2}{5} - \text{SSG} - \text{SS}(\text{Sc}) - \text{CM} = 10574.9$$

Source	df	SS	MS	F
G	1	6200.1	6200.100	2.09
Sc	3	246725.8	82241.933	27.75
G×Sc	3	10574.9	3524.967	1.19
Error	32	94825.6	2963.300	
Total	39	358326.4		

c The test statistic is $F = \text{MS}(\text{GSc})/\text{MSE} = 1.19$ and the rejection region is $F > 2.92$ (with $\alpha = .05$). Alternately, you can bound the p-value $> .10$. Hence, H_0 is not rejected. There is insufficient evidence to indicate interaction between gender and schools.

d You can see in the interaction plot that there is a small difference between the average scores for male and female students at schools 1 and 2, but no difference to speak of at the other two schools. The interaction is not significant.

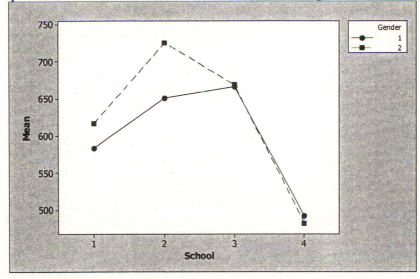

e The test statistic for testing gender is $F = 2.09$ with $F_{.05} = 4.17$ (or p-value $> .10$). The test statistic for schools is $F = 27.75$ with $F_{.05} = 2.92$ (or p-value $< .005$). There is a significant effect due to schools.

Using Tukey's method of paired comparisons with $\alpha = .01$, calculate

$$\omega = q_{.01}(4, 32)\frac{\sqrt{\text{MSE}}}{\sqrt{n_t}} = 4.80\sqrt{\frac{2963.3}{10}} = 82.63$$

The ranked means are shown below.

$$
\begin{array}{cccc}
487.1 & 600.1 & 667.4 & 688.6 \\
\bar{x}_4 & \bar{x}_1 & \bar{x}_3 & \bar{x}_2
\end{array}
$$

11.59   The objective is to determine whether or not mean reaction time differs for the five stimuli. The four people used in the experiment act as blocks, in an attempt to isolate the variation from person to person. A randomized block design is used, and the analysis of variance table is given in the printout.

a   The F statistic to detect a difference due to stimuli is

$$F = \frac{MST}{MSE} = 27.78$$

with p-value $= .000$. There is a significant difference in the effect of the five stimuli.

b   The treatment means can be further compared using Tukey's test with

$$\omega = q_{.05}(5,12)\frac{\sqrt{MSE}}{\sqrt{n_t}} = 4.51\sqrt{\frac{.00708}{4}} = .190$$

The ranked means are shown below.

E	A	B	D	C
.525	.7	.8	1.025	1.05

c   The F test for blocks produces $F = 6.59$ with p-value $= .007$. The block differences are significant; blocking has been effective.

11.63   This is similar to previous exercises. The complete ANOVA table is shown below.

Source	df	SS	MS	F
A	1	1.14	1.14	6.51
B	2	2.58	1.29	7.37
A×B	2	0.49	0.245	1.40
Error	24	4.20	0.175	
Total	29	8.41		

a   The test statistic is $F = MS(AB)/MSE = 1.40$ and the rejection region is $F > 3.40$. There is insufficient evidence to indicate an interaction.

b   Using Table 6 with $df_1 = 2$ and $df_2 = 24$, the following critical values are obtained.

α	.10	.05	.025	.01	.005
F_α	2.54	3.40	4.32	5.61	6.66

The observed value of F is less than $F_{.10}$, so that p-value $> .10$.

c   The test statistic for testing factor A is $F = 6.51$ with $F_{.05} = 4.26$. There is evidence that factor A affects the response.

d   The test statistic for factor B is $F = 7.37$ with $F_{.05} = 3.40$. Factor B also affects the response.

11.67   **a**   The design is a randomized block design, with weeks representing blocks and stores as treatments.

b   The *Minitab* computer printout is shown on the next page.

Two-way ANOVA: Total versus Week, Store

```
Source  DF      SS       MS      F       P
Week     3   571.71  190.570   8.27   0.003
Store    4   684.64  171.159   7.43   0.003
Error   12   276.38   23.032
Total   19  1532.73
S = 4.799    R-Sq = 81.97%    R-Sq(adj) = 71.45%
```

c The F test for treatments is $F = 7.43$ with p-value $= .003$. The p-value is small enough to allow rejection of H_0. There is a significant difference in the average weekly totals for the five supermarkets.

d With $k = 5$, $df = 12$, $n_t = 4$,

$$\omega = q_{.05}(5,12)\frac{\sqrt{MSE}}{\sqrt{n_t}} = 4.51\sqrt{\frac{23.032}{4}} = 10.82$$

The ranked means are shown below.

1	5	4	3	2
240.23	249.19	252.18	254.87	256.99

11.71   **a**   The design is a completely randomized design with three samples, each having a different number of measurements.

b   Use the computing formulas in Section 11.5 or the *Minitab* printout below.

One-way ANOVA: Iron versus Site

```
Source  DF      SS       MS       F       P
Site     2  132.277   66.139  126.85  0.000
Error   21   10.950    0.521
Total   23  143.227
S = 0.7221   R-Sq = 92.36%    R-Sq(adj) = 91.63%
```

The F test for treatments has a test statistic $F = 126.85$ with p-value $= .000$. The null hypothesis is rejected and we conclude that there is a significant difference in the average percentage of iron oxide at the three sites.

c   The diagnostic plots are shown on the next page. There appears to be no violation of the normality assumptions; the variances may be unequal, judging by the differing bar widths above and below the center line.

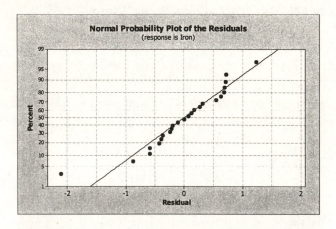

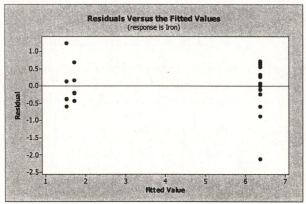

12: Linear Regression and Correlation

12.1 The line corresponding to the equation $y = 2x + 1$ can be graphed by locating the y values corresponding to $x = 0$, 1, and 2.

$$\text{When } x = 0, \ y = 2(0) + 1 = 1$$
$$\text{When } x = 1, \ y = 2(1) + 1 = 3$$
$$\text{When } x = 2, \ y = 2(2) + 1 = 5$$

The graph is shown below.

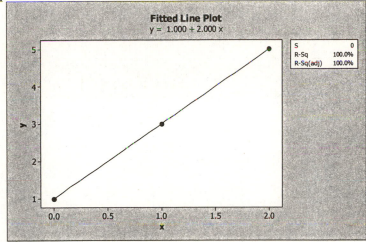

Note that the equation is in the form

$$y = \alpha + \beta x.$$

Thus, the slope of the line is $\beta = 2$ and the y-intercept is $\alpha = 1$.

12.5 A deterministic mathematical model is a model in which the value of a response y is exactly predicted from values of the variables that affect the response. On the other hand, a probabilistic mathematical model is one that contains random elements with specific probability distributions. The value of the response y in this model is not exactly determined.

12.9 **a** Calculate

$$\sum x_i = 850.8 \qquad \sum y_i = 3.755 \qquad \sum x_i y_i = 443.7727$$
$$\sum x_i^2 = 101,495.78 \qquad \sum y_i^2 = 1.941467 \qquad n = 9$$

Then

$$S_{xy} = \sum x_i y_i - \frac{(\sum x_i)(\sum y_i)}{n} = 88.80003333$$

$$S_{xx} = \sum x_i^2 - \frac{\left(\sum x_i\right)^2}{n} = 21,066.82$$

$$S_{yy} = \sum y_i^2 - \frac{\left(\sum y_i\right)^2}{n} = 0.3747976$$

b-c

$$b = \frac{S_{xy}}{S_{xx}} = \frac{88.800033}{21066.82} = 0.00421516$$

and $a = \bar{y} - b\bar{x} = 0.41722 - 0.00421516(94.5333) = 0.187$

The least squares line is $\hat{y} = a + bx = 0.187 + 0.0042x$. The graph of the least squares line and the nine data points are shown below.

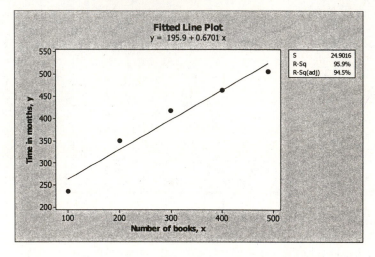

d   When $x = 100$, the value for y can be predicted using the least squares line as
$$\hat{y} = 0.187 + 0.0042(100) = 0.44$$

e   Calculate

$$SSR = \frac{\left(S_{xy}\right)^2}{S_{xx}} = \frac{(88.80003333)^2}{21,066.82} = .3743064$$

and

$$SSE = \text{Total SS} - SSR = S_{yy} - \frac{\left(S_{xy}\right)^2}{S_{xx}} = .3747976 - .374306417 = .00049118$$

The ANOVA table with 1 df for regression and $n - 2$ df for error is shown on the next page. Remember that the mean squares are calculated as $MS = SS/df$.

Source	df	SS	MS
Regression	1	.3743064	.374306
Error	7	.0004912	.0000070
Total	8	.3747976	

12.13 **a** The scatterplot generated by *Minitab* is shown below. The assumption of linearity is reasonable.

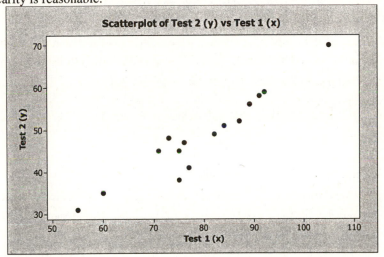

b Calculate

$$\sum x_i = 1192 \qquad \sum y_i = 725 \qquad \sum x_i y_i = 59,324$$

$$\sum x_i^2 = 96,990 \qquad \sum y_i^2 = 36,461 \qquad n = 15$$

Then

$$S_{xy} = \sum x_i y_i - \frac{\left(\sum x_i\right)\left(\sum y_i\right)}{n} = 1710.6667$$

$$S_{xx} = \sum x_i^2 - \frac{\left(\sum x_i\right)^2}{n} = 2265.7333$$

$$S_{yy} = \sum y_i^2 - \frac{\left(\sum y_i\right)^2}{n} = 1419.3333$$

$$b = \frac{S_{xy}}{S_{xx}} = \frac{1710.6667}{2265.7333} = .75502$$

and $a = \bar{y} - b\bar{x} = 48.3333 - (0.75502)(79.4667) = -11.665$
(using full accuracy). The least squares line is

$$\hat{y} = a + bx = -11.665 + 0.755x.$$

c When $x = 85$, the value for y can be predicted using the least squares line as

$$\hat{y} = a + bx = -11.665 + .755(85) = 52.51.$$

12.17 **a** The hypothesis to be tested is
$$H_0 : \beta = 0 \quad \text{versus} \quad H_a : \beta \neq 0$$

and the test statistic is a Student's t, calculated as
$$t = \frac{b - \beta_0}{\sqrt{MSE/S_{xx}}} = \frac{1.2 - 0}{\sqrt{0.533/10}} = 5.20$$

The critical value of t is based on $n - 2 = 3$ degrees of freedom and the rejection region for $\alpha = 0.05$ is $|t| > t_{.025} = 3.182$. Since the observed value of t falls in the rejection region, we reject H_0 and conclude that $\beta \neq 0$. That is, x is useful in the prediction of y.

b From the ANOVA table in Exercise 12.6, calculate
$$F = \frac{MSR}{MSE} = \frac{14.4}{0.5333} = 27.00$$

which is the square of the t statistic from part **a**: $t^2 = (5.20)^2 = 27.0$.

c The critical value of t from part **a** was $t_{.025} = 3.182$, while the critical value of F from part **b** with $df_1 = 1$ and $df_2 = 3$ is $F_{.05} = 10.13$. Notice that the relationship between the two critical values is
$$F = 10.13 = (3.182)^2 = t^2$$

12.21 **a** The dependent variable (to be predicted) is $y =$ cost and the independent variable is $x =$ distance.

b Preliminary calculations:

$$\sum x_i = 21,530 \qquad \sum y_i = 5052 \qquad \sum x_i y_i = 7,569,999$$
$$\sum x_i^2 = 37,763,314 \qquad \sum y_i^2 = 1,695,934 \qquad n = 18$$

Then
$$S_{xy} = \sum x_i y_i - \frac{(\sum x_i)(\sum y_i)}{n} = 1,527,245.667$$

$$S_{xx} = \sum x_i^2 - \frac{(\sum x_i)^2}{n} = 12,011,041.78$$

$$b = \frac{S_{xy}}{S_{xx}} = 0.127153$$

$$a = \bar{y} - b\bar{x} = 280.6667 - 0.127153(1196.1111) = 128.57699$$

and the least squares line is $\hat{y} = a + bx = 128.57699 + 0.127153x$.

c   The plot is shown below. The line appears to fit well through the 18 data points.

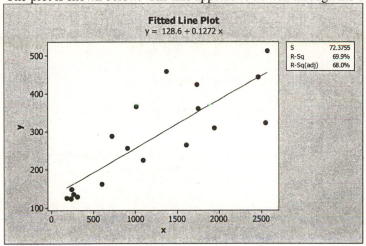

d   Calculate Total SS $= S_{yy} = \sum y_i^2 - \dfrac{(\sum y_i)^2}{n} = 1{,}695{,}934 - \dfrac{(5052)^2}{18} = 278{,}006.$

Then

$$\text{SSE} = S_{yy} - \frac{(S_{xy})^2}{S_{xx}} = 278{,}006 - \frac{(1{,}527{,}245.667)^2}{12{,}011{,}041.78} = 83{,}811.41055$$

and $\text{MSE} = \dfrac{\text{SSE}}{n-2} = \dfrac{83{,}811.41055}{16} = 5238.213.$ The hypothesis to be tested is

$$H_0 : \beta = 0 \quad \text{versus} \quad H_a : \beta \ne 0$$

and the test statistic is

$$t = \frac{b - \beta_0}{\sqrt{\text{MSE}/S_{xx}}} = \frac{0.127153 - 0}{\sqrt{5238.213/12{,}011{,}041.78}} = 6.09$$

The critical value of t is based on $n - 2 = 16$ degrees of freedom and the rejection region for $\alpha = 0.05$ is $|t| > t_{.025} = 2.120$, and H_0 is rejected. There is evidence at the 5% level to indicate that x and y are linearly related. That is, the regression model $y = \alpha + \beta x + \varepsilon$ is useful in predicting cost y.

12.25   **a**   The scatterplot generated by *Minitab* is shown on the next page. The assumption of linearity is reasonable.

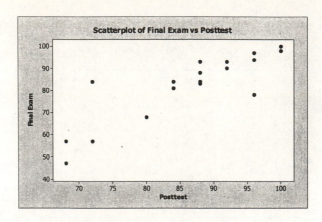

b   Using the *Minitab* printout, the equation of the regression line is $y = -26.82 + 1.2617x$.

c   The hypothesis to be tested is
$$H_0 : \beta = 0 \quad \text{versus} \quad H_a : \beta \neq 0$$

and the test statistic is a Student's t, calculated as
$$t = \frac{b - \beta_0}{\sqrt{\text{MSE}/S_{xx}}} = 7.49$$

with p-value $= .000$. Since the p-value is less than $\alpha = .01$, we reject H_0 and conclude that $\beta \neq 0$. That is, final exam scores and posttest scores are linearly related.

d   The 99% confidence interval for the slope β is
$$b \pm t_{\alpha/2}\sqrt{\text{MSE}/S_{xx}} \Rightarrow 1.2617 \pm 2.878(0.1253) \Rightarrow 1.2617 \pm 0.4849$$
or $0.7768 < \beta < 1.7466$.

12.29   Use a plot of residuals versus fits. The plot should appear as a random scatter of points, free of any patterns.

12.33   **a**   If you look carefully, there appears to be a slight curve to the five points.
    b   The fit of the regression line, measured as $r^2 = 0.959$ indicates that 95.9% of the overall variation can be explained by the straight line model.
    c   When we look at the residuals there is a strong curvilinear pattern that has not been explained by the straight line model. The relationship between time in months and number of books appears to be curvilinear.

12.39   **a**   Although very slight, the student might notice a slight curvature to the data points.
    b   The fit of the linear model is very good, assuming that this is *indeed* the correct model for this data set.

124

c   The normal probability plot follows the correct pattern for the assumption of normality. However, the residuals show the pattern of a quadratic curve, indicating that a quadratic rather than a linear model may have been the correct model for this data.

12.42   **a**   **a**   Use a computer program or the hand calculations shown below.

$$\sum x_i = 116 \qquad \sum y_i = 1480 \qquad \sum x_i y_i = 36,133$$

$$\sum x_i^2 = 2818 \qquad \sum y_i^2 = 467,600 \qquad n = 5$$

Then

$$S_{xy} = \sum x_i y_i - \frac{(\sum x_i)(\sum y_i)}{n} = 1797$$

$$S_{xx} = \sum x_i^2 - \frac{(\sum x_i)^2}{n} = 126.8$$

$$S_{yy} = \sum y_i^2 - \frac{(\sum y_i)^2}{n} = 29,520$$

$$b = \frac{S_{xy}}{S_{xx}} = \frac{1797}{126.8} = 14.17192$$

$$a = \bar{y} - b\bar{x} = 296 - 14.17192(23.2) = -32.789$$

and the least squares line is

$$\hat{y} = -32.789 + 14.172x.$$

b   The proportion of the total variation explained by regression is

$$r^2 = \frac{S_{xy}^2}{S_{xx}S_{yy}} = \frac{1797^2}{(126.8)(29,520)} = 0.8627$$

c   The diagnostic plots, generated by *Minitab* are shown below. The plots do not show any strong violation of assumptions.

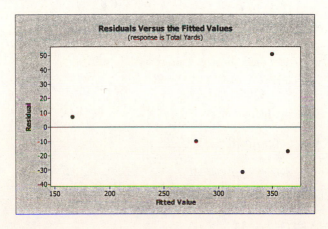

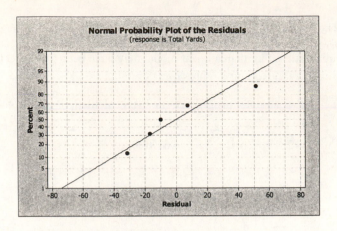

12.43   Refer to Exercise 12.42 and calculate

$$\text{SSE} = S_{yy} - \frac{\left(S_{xy}\right)^2}{S_{xx}} = 29520 - \frac{1797^2}{126.8} = 4053.05205$$

and $\text{MSE} = \dfrac{\text{SSE}}{n-2} = \dfrac{4053.05205}{3} = 1351.01735.$

a   The point estimator for $E(y)$ when $x = 21$ is
$$\hat{y} = -32.789 + 14.172(21) = 264.823$$

and the 95% confidence interval is

$$\hat{y} \pm t_{.025}\sqrt{\text{MSE}\left(\frac{1}{n} + \frac{(x_p - \overline{x})^2}{S_{xx}}\right)}$$

$$264.823 \pm 3.182\sqrt{1351.01735\left(\frac{1}{5} + \frac{(21-23.2)^2}{126.8}\right)}$$

$$264.823 \pm 57.079$$

or $207.744 < E(y) < 321.902$.

b   The point estimator for y when $x = 21$ is still
$$\hat{y} = -32.789 + 14.172(21) = 264.823$$

and the 95% prediction interval is

$$\hat{y} \pm t_{.025}\sqrt{\text{MSE}\left(1 + \frac{1}{n} + \frac{(x_p - \overline{x})^2}{S_{xx}}\right)}$$

$$264.823 \pm 3.182\sqrt{1351.01735\left(1 + \frac{1}{5} + \frac{(21-23.2)^2}{126.8}\right)}$$

$$264.823 \pm 130.143$$

or $134.680 < y < 394.966$.

c   This would not be advisable, since you are trying to estimate outside the range of experimentation.

12.47   **a**   Refer to figure below. The sample correlation coefficient will be positive.

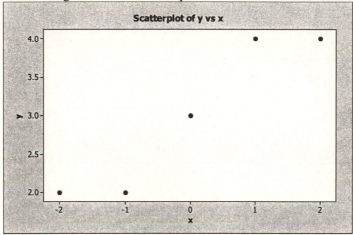

b   Calculate

$$S_{xy} = \sum x_i y_i - \frac{(\sum x_i)(\sum y_i)}{n} = 6 - \frac{0(15)}{5} = 6$$

$$S_{xx} = \sum x_i^2 - \frac{(\sum x_i)^2}{n} = 10 - \frac{0^2}{5} = 10$$

$$S_{yy} = \sum y_i^2 - \frac{(\sum y_i)^2}{n} = 49 - \frac{15^2}{5} = 4$$

Then $r = \dfrac{S_{xy}}{\sqrt{S_{xx}S_{yy}}} = \dfrac{6}{\sqrt{40}} = 0.9487$ and $r^2 = (0.9487)^2 = 0.9000$. Approximately 90% of the total sum of squares of deviations was reduced by using the least squares equation instead of $\bar{y}$ as a predictor of y.

12.51   When the pre-test score x is high, the post-test score y should also be high. There should be a positive correlation.

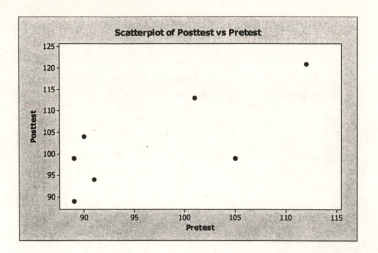

Calculate

$$S_{xy} = \sum x_i y_i - \frac{(\sum x_i)(\sum y_i)}{n} = 70,006 - \frac{677(719)}{7} = 468.42857$$

$$S_{xx} = \sum x_i^2 - \frac{(\sum x_i)^2}{n} = 65,993 - \frac{677^2}{7} = 517.42857$$

$$S_{yy} = \sum y_i^2 - \frac{(\sum y_i)^2}{n} = 74,585 - \frac{719^2}{7} = 733.42857$$

Then $r = \dfrac{S_{xy}}{\sqrt{S_{xx}S_{yy}}} = \dfrac{468.42857}{\sqrt{517.42857(733.42857)}} = 0.760$.

The test of hypothesis is

$$H_0 : \rho = 0 \quad \text{versus} \quad H_a : \rho > 0$$

and the test statistic is

$$t = \frac{r\sqrt{n-2}}{\sqrt{1-r^2}} = \frac{0.760\sqrt{5}}{\sqrt{1-(0.760)^2}} = 2.615$$

The rejection region for $\alpha = 0.05$ is $t > t_{.05} = 2.015$ and H_0 is rejected. There is sufficient evidence to indicate positive correlation.

12.55 **a**   Since neither of the two variables, amount of sodium or number of calories, is controlled, the methods of correlation rather than linear regression analysis should be used.

b   Use a computer program, your scientific calculator or the computing formulas given in the text to calculate the correlation coefficient r. The *Minitab* printout for this data set is shown on the next page.

Correlations: Sodium, Calories

```
Pearson correlation of Sodium and Calories = 0.981
P-Value = 0.003
```

There is evidence of a highly significant correlation, since the p-value is so small. The correlation is positive.

12.61   Answers will vary. The *Minitab* output for this linear regression problem is shown below.

Regression Analysis: y versus x

```
The regression equation is
y = 21.9 + 15.0 x

Predictor     Coef   SE Coef       T      P
Constant    21.867     3.502    6.24  0.000
x          14.9667    0.9530   15.70  0.000

S = 3.69098    R-Sq = 96.1%    R-Sq(adj) = 95.7%

Analysis of Variance

Source          DF       SS      MS        F      P
Regression       1   3360.0  3360.0   246.64  0.000
Residual Error  10    136.2    13.6
Total           11   3496.2
```

Correlations: x, y

```
Pearson correlation of x and y = 0.980
P-Value = 0.000
```

a   The correlation coefficient is $r = 0.980$.

b   The coefficient of determination is $r^2 = 0.961$ (or 96.1%).

c   The least squares line is $\hat{y} = 21.867 + 14.9667x$.

d   We wish to estimate the mean percentage of kill for an application of 4 pounds of nematicide per acre. Since the percent kill y is actually a binomial percentage, the variance of y will change depending on the value of p, the proportion of nematodes killed for a particular application rate. The residual plot versus the fitted values shows this phenomenon as a "football-shaped" pattern. The normal probability plot also shows some deviation from normality in the tails of the plot. A transformation may be needed to assure that the regression assumptions are satisfied.

12.65   **a**   Use a computer program, your scientific calculator or the computing formulas given in the text to calculate the correlation coefficient r.

$$S_{xy} = \sum x_i y_i - \frac{(\sum x_i)(\sum y_i)}{n} = 1,233,987 - \frac{5028(2856)}{12} = 37,323$$

$$S_{xx} = \sum x_i^2 - \frac{(\sum x_i)^2}{n} = 2,212,178 - \frac{5028^2}{12} = 105,446$$

$$S_{yy} = \sum y_i^2 - \frac{(\sum y_i)^2}{n} = 723,882 - \frac{2856^2}{12} = 44,154$$

129

Then $r = \dfrac{S_{xy}}{\sqrt{S_{xx}S_{yy}}} = \dfrac{37,323}{\sqrt{105,446(44,154)}} = 0.5470$.

The test of hypothesis is

$$H_0 : \rho = 0 \quad \text{versus} \quad H_a : \rho > 0$$

and the test statistic is

$$t = \frac{r\sqrt{n-2}}{\sqrt{1-r^2}} = \frac{0.5470\sqrt{10}}{\sqrt{1-(0.5470)^2}} = 2.066$$

with p-value $= P(t > 2.066)$ bounded as

$$0.05 < p\text{-value} < 0.10$$

If the experimenter is willing to tolerate a p-value this large, then H_0 can be rejected. Otherwise, you would declare the results not significant; there is insufficient evidence to indicate that bending stiffness and twisting stiffness are positively correlated.

b $r^2 = (0.5470)^2 = 0.2992$ so that 29.9% of the total variation in y can be explained by the independent variable x.

12.69   **a**   The plot is shown below. Notice that the relationship is fairly weak.

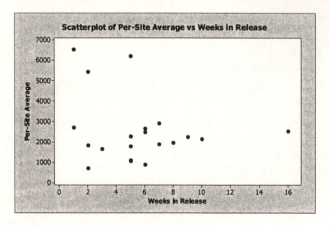

Regression Analysis: y versus x

```
The regression equation is
y = 3091 - 98 x

Predictor     Coef   SE Coef        T       P
Constant    3091.4     701.0     4.41   0.000
x            -97.7     107.2    -0.91   0.374
S = 1657.23    R-Sq = 4.4%    R-Sq(adj) = 0.0%

Analysis of Variance
Source           DF         SS        MS      F      P
Regression        1    2282896   2282896   0.83  0.374
Residual Error   18   49435210   2746401
Total            19   51718106
```

```
Unusual Observations
Obs     x       y     Fit   SE Fit   Residual   St Resid
  1    1.0    6551   2994     613       3557       2.31R
 10   16.0    2538   1527    1180       1011       0.87 X
 13    5.0    6219   2603     375       3616       2.24R
```

b   From the printout, $r^2 = 0.044$. Only about 4% of the overall variation in y is explained by using the linear model.

c   From the printout, the regression equation is $\hat{y} = 3091.4 - 97.7x$ and the regression is not significant $(t = -0.91$ with p-value $= 0.374)$.

d   Since the regression is not significant, it is not appropriate to use the regression line for estimation or prediction.

12.73   **a**   The calculations shown below are done using the computing formulas. An appropriate computer program will provide identical results to within rounding error.

$$\sum x_i = 150 \qquad\qquad \sum y_i = 91 \qquad\qquad \sum x_i y_i = 986$$
$$\sum x_i^2 = 2750 \qquad\qquad \sum y_i^2 = 1120.04 \qquad\qquad n = 10$$

Then

$$S_{xy} = \sum x_i y_i - \frac{(\sum x_i)(\sum y_i)}{n} = 986 - \frac{150(91)}{10} = -379$$

$$S_{xx} = \sum x_i^2 - \frac{(\sum x_i)^2}{n} = 2750 - \frac{150^2}{10} = 500$$

$$S_{yy} = \sum y_i^2 - \frac{(\sum y_i)^2}{n} = 1120.04 - \frac{91^2}{10} = 291.94$$

a

$$b = \frac{S_{xy}}{S_{xx}} = \frac{-379}{500} = -.758$$

$$a = \bar{y} - b\bar{x} = 9.1 - (-.758)(15) = 20.47$$

and the least squares line is $\hat{y} = a + bx = 20.47 - .758x$.

b   Since Total SS $= S_{yy} = 291.94$ and

$$SSR = \frac{(S_{xy})^2}{S_{xx}} = \frac{(-379)^2}{500} = 287.282$$

Then $SSE = \text{Total SS} - SSR = S_{yy} - \frac{(S_{xy})^2}{S_{xx}} = 4.658$

The ANOVA table with 1 df for regression and $n - 2$ df for error is shown below. Remember that the mean squares are calculated as $MS = SS/df$.

Source	df	SS	MS
Regression	1	287.282	287.282
Error	8	4.658	.58225
Total	9	291.940	

131

c To test $H_0 : \beta = 0, H_a : \beta \neq 0$, the test statistic is

$$t = \frac{b - \beta_0}{s/\sqrt{S_{xx}}} = \frac{-.758}{\sqrt{.58225/500}} = -22.21$$

The rejection region for $\alpha = 0.05$ is $|t| > t_{.025} = 2.306$ and we reject H_0. There is sufficient evidence to indicate that x and y are linearly related.

d The 95% confidence interval for the slope β is

$$b \pm t_{\alpha/2}\sqrt{MSE/S_{xx}} \Rightarrow -.758 \pm 2.896\sqrt{.58225/500} \Rightarrow -.758 \pm .099$$

or $-.857 < \beta < -.659$.

e When $x = 14$, the estimate of expected freshness $E(y)$ is
$\hat{y} = 20.47 - .758(14) = 9.858$ and the 95% confidence interval is

$$\hat{y} \pm t_{.025}\sqrt{MSE\left(\frac{1}{n} + \frac{(x_p - \bar{x})^2}{S_{xx}}\right)}$$

$$9.858 \pm 2.306\sqrt{.58225\left(\frac{1}{10} + \frac{(14-15)^2}{500}\right)}$$

$9.858 \pm .562$

or $9.296 < E(y) < 10.420$.

f Calculate

$$r^2 = \frac{SSR}{\text{Total SS}} = \frac{287.282}{291.94} = 0.984$$

The total variation has been reduced by 98.4%% by using the linear model.

12.76-77 Use the **How a Line Works** applet. The line $y = 0.5x + 3$ has a slope of 0.5 and a y-intercept of 3, while the line $y = -0.5x + 3$ has a slope of -0.5 and a y-intercept of 3. The second line slopes downward at the same rate as the first line slopes upward. They both cross the y axis at the same point.

12.80 **a** Use a computer, your scientific calculator or the computing formulas to find the correlation between x and y. The *Minitab* correlation printout below shows $r = 0.231$ with p-value = 0.549 which is not significant at the 5% level of significance. You cannot conclude that there is a significant positive correlation between median rate and score for "budget" hotels.

Correlations: Median Rate, Score
```
Pearson correlation of Median Rate and Score = 0.231
P-Value = 0.549
```

b-c   Use the **Correlation and the Scatterplot** applet. There is a random scatter of points, with no outliers. The student's plot should look similar to the *Minitab* plot shown below.

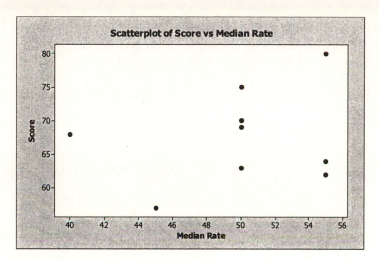

13: Multiple Regression Analysis

13.1 **a** When $x_2 = 2$, $E(y) = 3 + x_1 - 2(2) = x_1 - 1$.

When $x_2 = 1$, $E(y) = 3 + x_1 - 2(1) = x_1 + 1$.

When $x_2 = 0$, $E(y) = 3 + x_1 - 2(0) = x_1 + 3$.

These three straight lines are graphed below.

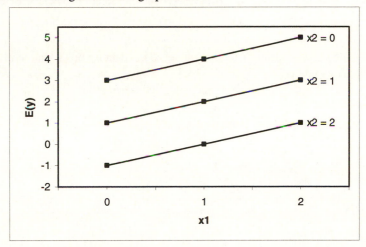

b Notice that the lines are parallel (they have the same slope).

13.5 **a** The model is quadratic.

b Since $R^2 = .815$, the sum of squares of deviations is reduced by 81.5% using the quadratic model rather than $\bar{y}$ to predict y.

c The hypothesis to be tested is

$H_0 : \beta_1 = \beta_2 = 0$ H_a: at least one β_i differs from zero

and the test statistic is

$$F = \frac{MSR}{MSE} = 37.37$$

which has an F distribution with $df_1 = k = 2$ and $df_2 = n - k - 1 = 20 - 2 - 1 = 17$. The p-value given in the printout is P = .000 and H_0 is rejected. There is evidence that the model contributes information for the prediction of y.

13.9 **a** Rate of increase is measured by the slope of a line tangent to the curve; this line is given by an equation obtained as dy/dx, the derivative of y with respect to x. In particular,

$$\frac{dy}{dx} = \frac{d}{dx}\left(\beta_0 + \beta_1 x + \beta_2 x^2\right) = \beta_1 + 2\beta_2 x$$

which has slope $2\beta_2$. If β_2 is negative, then the rate of increase is decreasing. Hence, the hypothesis of interest is

$$H_0 : \beta_2 = 0, \quad H_a : \beta_2 < 0$$

b The individual t-test is $t = -8.11$ as in Exercise 13.8b. However, the test is one-tailed, which means that the p-value is half of the amount given in the printout. That is, p-value $= \frac{1}{2}(.000) = .000$. Hence, H_0 is again rejected. There is evidence to indicate a decreasing rate of increase.

13.15 **a** The *Minitab* printout fitting the model to the data is shown on the next page. The least squares line is

$$\hat{y} = -8.177 + 0.292x_1 + 4.434x_2$$

Regression Analysis: y versus x1, x2
```
The regression equation is
y = - 8.18 + 0.292 x1 + 4.43 x2

Predictor     Coef   SE Coef       T      P
Constant    -8.177     4.206   -1.94  0.093
x1          0.2921    0.1357    2.15  0.068
x2          4.4343    0.8002    5.54  0.001
S = 3.30335   R-Sq = 82.3%   R-Sq(adj) = 77.2%

Analysis of Variance
Source          DF       SS       MS       F      P
Regression       2   355.22   177.61   16.28  0.002
Residual Error   7    76.38    10.91
Total            9   431.60

Source   DF   Seq SS
x1        1    20.16
x2        1   335.05
```

b The F test for the overall utility of the model is $F = 16.28$ with $P = .002$. The results are highly significant; the model contributes significant information for the prediction of y.

c To test the effect of advertising expenditure, the hypothesis of interest is

$$H_0 : \beta_2 = 0, \quad H_a : \beta_2 \neq 0$$

and the test statistic is $t = 5.54$ with p-value $= .001$. Since $\alpha = .01$, H_0 is rejected. We conclude that advertising expenditure contributes significant information for the prediction of y, given that capital investment is already in the model.

d From the *Minitab* printout, R-Sq = 82.3%, which means that 82.3% of the total variation can be explained by the quadratic model. The model is very effective.

13.19 **a** The variable x_2 must be the quantitative variable, since it appears as a quadratic term in the model. Qualitative variables appear only with exponent 1, although they may appear as the coefficient of another quantitative variable with exponent 2 or greater.

b When $x_1 = 0$, $\hat{y} = 12.6 + 3.9x_2^2$ while when $x_1 = 1$,

$$\hat{y} = 12.6 + .54(1) - 1.2x_2 + 3.9x_2^2$$
$$= 13.14 - 1.2x_2 + 3.9x_2^2$$

c The graph below shows the two parabolas.

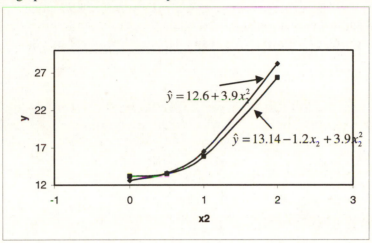

13.23 The basic response equation for a specific type of bonding compound would be

$$E(y) = \beta_0 + \beta_1 x_1 + \beta_2 x_1^2$$

Since the qualitative variable "bonding compound" is at two levels, one dummy variable is needed to incorporate this variable into the model. Define the dummy variable x_2 as follows:

$$x_2 = 1 \text{ if bonding compound 2}$$
$$= 0 \text{ otherwise}$$

The expanded model is now written as

$$E(y) = \beta_0 + \beta_1 x_1 + \beta_2 x_1^2 + \beta_3 x_2 + \beta_4 x_1 x_2 + \beta_5 x_1^2 x_2$$

13.25 **a** From the printout, the prediction equation is $\hat{y} = 8.585 + 3.8208x - 0.21663x^2$.

b R^2 is labeled "R-sq" or $R^2 = .944$. Hence 94.4% of the total variation is accounted for by using x and x^2 in the model.

c The hypothesis of interest is

$$H_0 : \beta_1 = \beta_2 = 0 \qquad H_a: \text{at least one } \beta_i \text{ differs from zero}$$

and the test statistic is $F = 33.44$ with p-value $= .003$. Hence, H_0 is rejected, and we conclude that the model contributes significant information for the prediction of y.

d The hypothesis of interest is

$$H_0 : \beta_2 = 0 \qquad H_a: \beta_2 \neq 0$$

and the test statistic is $t = -4.93$ with p-value $= .008$. Hence, H_0 is rejected, and we conclude that the quadratic model provides a better fit to the data than a simple linear model.

e The pattern of the diagnostic plots does not indicate any obvious violation of the regression assumptions.

13.29 **a** The model is

$$y = \beta_0 + \beta_1 x_1 + \beta_2 x_2 + \beta_3 x_1^2 + \beta_4 x_1 x_2 + \beta_5 x_1^2 x_2 + \varepsilon$$

and the Minitab printout is shown below.

Regression Analysis: y versus x1, x2, x1sq, x1x2, x1sqx2
```
The regression equation is
y = 4.5 + 6.39 x1 - 50.9 x2 + 0.132 x1sq + 17.1 x1x2 - 0.502 x1sqx2

Predictor      Coef   SE Coef       T      P
Constant       4.51     42.24    0.11  0.916
x1            6.394     5.777    1.11  0.275
x2           -50.85     56.21   -0.90  0.371
x1sq         0.1318    0.1687    0.78  0.439
x1x2         17.064     7.101    2.40  0.021
x1sqx2      -0.5025    0.1992   -2.52  0.016
S = 71.6891   R-Sq = 76.8%   R-Sq(adj) = 73.8%

Analysis of Variance
Source           DF       SS      MS      F      P
Regression        5   664164  132833  25.85  0.000
Residual Error   39   200434    5139
Total            44   864598
```

b The fitted prediction model uses the coefficients given in the column marked "Coef" in the printout:

$$\hat{y} = 4.51 + 6.394 x_1 - 50.85 x_2 + 17.064 x_1 x_2 + .1318 x_1^2 - .5025 x_1^2 x_2$$

The F test for the model's utility is $F = 25.85$ with $P = .000$ and $R^2 = .768$. The model fits quite well.

c If the dolphin is female, $x_2 = 0$ and the prediction equation becomes

$$\hat{y} = 4.51 + 6.394 x_1 + .1318 x_1^2$$

d If the dolphin is male, $x_2 = 1$ and the prediction equation becomes

$$\hat{y} = -46.34 + 23.458 x_1 - .3707 x_1^2$$

e The hypothesis of interest is

$$H_0 : \beta_4 = 0 \qquad H_a : \beta_4 \neq 0$$

and the test statistic is $t = .78$ with p-value $= .439$. H_0 is not rejected and we conclude that the quadratic term is not important in predicting mercury concentration for female dolphins.

13.31 **a-b** The data is plotted on the next page. It appears to be a curvilinear relationship, which could be described using the quadratic model $y = \beta_0 + \beta_1 x + \beta_2 x^2 + \varepsilon$.

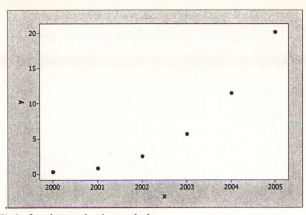

c The *Minitab* printout is shown below.

Regression Analysis: y versus x, x_sq

```
The regression equation is
y = 4114749 - 4113 x + 1.03 x_sq

Predictor       Coef   SE Coef        T       P
Constant     4114749    343582    11.98   0.001
x            -4113.4     343.2   -11.99   0.001
x_sq         1.02804   0.08568    12.00   0.001
S = 0.523521   R-Sq = 99.7%   R-Sq(adj) = 99.5%

Analysis of Variance
Source          DF       SS       MS       F       P
Regression       2   297.16   148.58   542.11   0.000
Residual Error   3     0.82     0.27
Total            5   297.98
```

d The hypothesis of interest is

$$H_0 : \beta_1 = \beta_2 = 0$$

and the test statistic is $F = 542.11$ with p-value $= .000$. H_0 is rejected and we conclude that the model provides valuable information for the prediction of y.

e $R^2 = .997$. Hence, 99.7% of the total variation is accounted for by using x and x^2 in the model.

f The residual plots are shown below. There is no reason to doubt the validity of the regression assumptions.

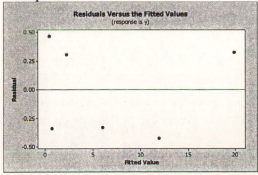

139

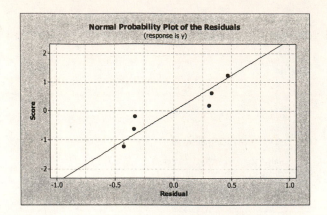

13.35   **a** $R^2 = .999$. Hence, 99.9% of the total variation is accounted for by using x and x^2 in the model.

b   The hypothesis of interest is
$$H_0 : \beta_1 = \beta_2 = 0$$
and the test statistic is $F = 1676.61$ with p-value $= .000$. H_0 is rejected and we conclude that the model provides valuable information for the prediction of y.

c   The hypothesis of interest is
$$H_0 : \beta_1 = 0$$
and the test statistic is $t = -2.65$ with p-value $= .045$. H_0 is rejected and we conclude that the linear regression coefficient is significant when x^2 is in the model.

d   The hypothesis of interest is
$$H_0 : \beta_2 = 0$$
and the test statistic is $t = 15.14$ with p-value $= .000$. H_0 is rejected and we conclude that the quadratic regression coefficient is significant when x is in the model.

e   When the quadratic term is removed from the model, the value of R^2 decreases by $99.9 - 93.0 = 6.9\%$. This is the additional contribution of the quadratic term when it is included in the model.

f   The clear pattern of a curve in the residual plot indicates that the quadratic term should be included in the model.

14: Analysis of Categorical Data

14.3 For a test of specified cell probabilities, the degrees of freedom are $k-1$. Use Table 5, Appendix I:

a $df = 6$; $\chi^2_{.05} = 12.59$; reject H_0 if $X^2 > 12.59$

b $df = 9$; $\chi^2_{.01} = 21.666$; reject H_0 if $X^2 > 21.666$

c $df = 13$; $\chi^2_{.005} = 29.814$; reject H_0 if $X^2 > 29.8194$

d $df = 2$; $\chi^2_{.05} = 5.99$; reject H_0 if $X^2 > 5.99$

14.7 One thousand cars were each classified according to the lane which they occupied (one through four). If no lane is preferred over another, the probability that a car will be driven in lane i, $i = 1, 2, 3, 4$ is ¼. The null hypothesis is then

$$H_0 : p_1 = p_2 = p_3 = p_4 = \frac{1}{4}$$

and the test statistic is

$$X^2 = \sum \frac{(O_i - E_i)^2}{E_i}$$

with $E_i = np_i = 1000(1/4) = 250$ for $i = 1, 2, 3, 4$. A table of observed and expected cell counts follows:

Lane	1	2	3	4
O_i	294	276	238	192
E_i	250	250	250	250

Then

$$X^2 = \frac{(294-250)^2}{250} + \frac{(276-250)^2}{250} + \frac{(238-250)^2}{250} + \frac{(192-250)^2}{250}$$

$$= \frac{6120}{250} = 24.48$$

The rejection region with $k-1 = 3$ df is $X^2 > \chi^2_{.05} = 7.81$. Since the observed value of X^2 falls in the rejection region, we reject H_0. There is a difference in preference for the four lanes.

14.11 Similar to previous exercises. The hypothesis to be tested is

$$H_0 : p_1 = p_2 = \cdots = p_{12} = \frac{1}{12}$$

versus $\quad H_a$: at least one p_i is different from the others

with

$$E_i = np_i = 400(1/12) = 33.333 .$$

The test statistic is

$$X^2 = \frac{(38-33.33)^2}{33.33} + \cdots + \frac{(35-33.33)^2}{33.33} = 13.58$$

The upper tailed rejection region is with $\alpha = .05$ and $k-1 = 11$ df is $X^2 > \chi^2_{.05} = 19.675$. The null hypothesis is not rejected and we cannot conclude that the proportion of cases varies from month to month.

14.15 It is necessary to determine whether admission rates differ from the previously reported rates. A table of observed and expected cell counts follows:

	Unconditional	Trial	Refused	Totals
O_i	329	43	128	500
E_i	300	25	175	500

The null hypothesis to be tested is
$$H_0 : p_1 = .60; \ p_2 = .05; \ p_3 = .35$$
against the alternative that at least one of these probabilities is incorrect. The test statistic is

$$X^2 = \frac{(329-300)^2}{300} + \frac{(43-25)^2}{25} + \frac{(128-175)^2}{175} = 28.386$$

The number of degrees of freedom is $k-1 = 2$ and the rejection region $X^2 > \chi^2_{.05} = 5.99$. The null hypothesis is rejected, and we conclude that there has been a departure from previous admission rates. Notice that the percentage of unconditional admissions has risen slightly, the number of conditional admissions has increased, and the percentage refused admission has decreased at the expense of the fist two categories.

14.17 Refer to Section 14.4 of the text. For a 3×5 contingency table with $r = 3$ and $c = 5$, there are $(r-1)(c-1) = (2)(4) = 8$ degrees of freedom.

14.21 **a** The hypothesis of independence between attachment pattern and child care time is tested using the chi-square statistic. The contingency table, including column and row totals and the estimated expected cell counts, follows.

	Child Care			
Attachment	Low	Moderate	High	**Total**
Secure	24	35	5	64
	(24.09)	(30.97)	(8.95)	
Anxious	11	10	8	29
	(10.91)	(14.03)	(4.05)	
Total	111	51	297	459

The test statistic is
$$X^2 = \frac{(24-24.09)^2}{24.09} + \frac{(35-30.97)^2}{30.97} + \cdots + \frac{(8-4.05)^2}{4.05} = 7.267$$

and the rejection region is $X^2 > \chi^2_{.05} = 5.99$ with 2 df. H_0 is rejected. There is evidence of a dependence between attachment pattern and child care time.

b The value $X^2 = 7.267$ is between $\chi^2_{.05}$ and $\chi^2_{.025}$ so that $.025 < p\text{-value} < .05$. The results are significant.

14.25 **a** The hypothesis of independence between salary and number of workdays at home is tested using the chi-square statistic. The contingency table, including column and row totals and the estimated expected cell counts, generated by *Minitab* follows.

Chi-Square Test: Less than one, At least one, not all, All at home
Expected counts are printed below observed counts
Chi-Square contributions are printed below expected counts

	Less than one	At least one, not all	All at home	Total
1	38	16	14	68
	36.27	21.08	10.65	
	0.083	1.224	1.051	
2	54	26	12	92
	49.07	28.52	14.41	
	0.496	0.223	0.404	
3	35	22	9	66
	35.20	20.46	10.34	
	0.001	0.116	0.174	
4	33	29	12	74
	39.47	22.94	11.59	
	1.060	1.601	0.014	
Total	160	93	47	300

Chi-Sq = 6.447, DF = 6, P-Value = 0.375
The test statistic is

$$X^2 = \frac{(38-36.27)^2}{36.27} + \frac{(16-21.08)^2}{21.08} + \cdots + \frac{(12-11.59)^2}{11.59} = 6.447$$

and the rejection region with $\alpha = .05$ and $df = 3(2) = 6$ is $X^2 > \chi^2_{.05} = 12.59$ and the null hypothesis is not rejected. There is insufficient evidence to indicate that salary is dependent on the number of workdays spent at home.

b The observed value of the test statistic, $X^2 = 6.447$, is less than $\chi^2_{.10} = 10.6446$ so that the p-value is more than .10. This would confirm the non-rejection of the null hypothesis from part a.

14.29 Because a set number of Americans in each sub-population were each fixed at 200, we have a contingency table with fixed rows. The table, with estimated expected cell counts appearing in parentheses, is shown on the next page.

	Yes	No	Total
White-American	40 (62)	160 (138)	200
African-American	56 (62)	144 (138)	200
Hispanic-American	68 (62)	132 (138)	200
Asian-American	84 (62)	116 (138)	200
Total	248	552	800

The test statistic is

$$X^2 = \frac{(40-62)^2}{62} + \frac{(56-62)^2}{62} + \cdots + \frac{(116-138)^2}{138} = 24.31$$

and the rejection region with 3 df is $X^2 > 11.3449$. H_0 is rejected and we conclude that the incidence of parental support is dependent on the sub-population of Americans.

14.33 The number of observations per column were selected prior to the experiment. The test procedure is identical to that used for an $r \times c$ contingency table. The contingency table, including column and row totals and the estimated expected cell counts, follows.

Family Members	Type			Total
	Apartment	Duplex	Single Residence	
1	8 (9.67)	20 (9.67)	1 (9.67)	29
2	16 (11)	8 (11)	9 (11)	33
3	10 (11.33)	10 (11.33)	14 (11.33)	34
4 or more	6 (8)	2 (8)	16 (8)	24
Total	40	40	40	120

The test statistic is

$$X^2 = \frac{(8-9.67)^2}{9.67} + \frac{(20-9.67)^2}{9.67} + \cdots + \frac{(16-8)^2}{8} = 36.499$$

using computer accuracy. With $(r-1)(c-1) = 6$ df and $\alpha = .01$, the rejection region is $X^2 > 16.8119$. The null hypothesis is rejected. There is sufficient evidence to indicate that family size is dependent on type of family residence. It appears that as the family size increases, it is more likely that people will live in single residences.

14.35 If the housekeeper actually has no preference, he or she has an equal chance of picking any of the five floor polishes. Hence, the null hypothesis to be tested is

$$H_0 : p_1 = p_2 = p_3 = p_4 = p_5 = \frac{1}{5}$$

The values of O_i are the actual counts observed in the experiment, and
$E_i = np_i = 100(1/5) = 20$.

Polish	A	B	C	D	E
O_i	27	17	15	22	19
E_i	20	20	20	20	20

Then
$$X^2 = \frac{(27-20)^2}{20} + \frac{(17-20)^2}{20} + \cdots + \frac{(19-20)^2}{20} = 4.40$$

The p-value with $df = k - 1 = 4$ is greater than .10 and H_0 is not rejected. We cannot conclude that there is a difference in the preference for the five floor polishes. Even if this hypothesis **had** been rejected, the conclusion would be that at least one of the value of the p_i was significantly different from 1/6. However, this does not imply that p_i is necessarily greater than 1/6. Hence, we could not conclude that polish A is superior.

If the objective of the experiment is to show that polish A is superior, a better procedure would be to test an hypothesis as follows:
$$H_0 : p_1 = 1/6 \qquad H_a : p_1 > 1/6$$
From a sample of $n = 100$ housewives, $x = 27$ are found to prefer polish A. A z-test can be performed on the single binomial parameter p_1.

14.39 **a** To test for homogeneity of the five binomial populations, we use chi-square statistic and the 5×2 contingency table shown below The null hypothesis is that voter choice and church attendance are independent, with p be the proportion of voters who intend to vote for G.W. Bush in the 2004 election for a particular church attendance group. The contingency table generated by *Minitab* is shown below

Chi-Square Test: G.W. Bush, Democrat

Expected counts are printed below observed counts
Chi-Square contributions are printed below expected counts

	G.W. Bush	Democrat	Total
1	89	53	142
	73.64	68.36	
	3.205	3.453	
2	87	68	155
	80.38	74.62	
	0.546	0.588	
3	93	85	178
	92.30	85.70	
	0.005	0.006	
4	114	134	248
	128.60	119.40	
	1.658	1.786	

| Total | 405 | 376 | 781 |

```
Chi-Sq = 15.752, DF = 4, P-Value = 0.003
```

The observed value of the test statistic is $X^2 = 15.752$ with p-value $= .003$ and the null hypothesis is rejected at the 5% level of significance. There is sufficient evidence to indicate that the proportion of adults who intend to vote for G.W. Bush in the 2004 election is dependent on church attendance.

14.43 The flower fall into one of four classifications, with theoretical ratio 9:3:3:1. Converting these ratios to probabilities,

$$p_1 = 9/16 = .5625 \qquad p_2 = 3/16 = .1875$$
$$p_3 = 3/16 = .1875 \qquad p_4 = 1/16 = .0625$$

We will test the null hypothesis that the probabilities are as above against the alternative that they differ. The table of observed and expected cell counts follows:

	AB	Ab	aB	aa
O_i	95	30	28	7
E_i	90	30	30	10

The test statistic is

$$X^2 = \frac{(95-90)^2}{90} + \frac{(30-30)^2}{30} + \frac{(28-30)^2}{30} + \frac{(7-10)^2}{10} = 1.311$$

The number of degrees of freedom is $k-1=3$ and the rejection region with $\alpha = .01$ is $X^2 > \chi^2_{.01} = 11.3449$. Since the observed value of X^2 does not fall in the rejection region, we do not reject H_0. We do not have enough information to contradict the theoretical model for the classification of flower color and shape.

14.48 In order to perform a chi-square "goodness of fit" test on the given data, it is necessary that the values O_i and E_i are known for each of the five cells. The O_i (the number of measurements falling in the i-th cell) are given. However, $E_i = np_i$ must be calculated. Remember that p_i is the probability that a measurement falls in the i-th cell. The hypothesis to be tested is

H_0 : the experiment is binomial   versus H_a : the experiment is not binomial

Let x be the number of successes and p be the probability of success on a single trial. Then, assuming the null hypothesis to be true,

$$p_0 = P(x=0) = C_0^4 p^0 (1-p)^4 \qquad p_1 = P(x=1) = C_1^4 p^1 (1-p)^3$$
$$p_2 = P(x=2) = C_2^4 p^2 (1-p)^2 \qquad p_3 = P(x=3) = C_3^4 p^3 (1-p)^1$$
$$p_4 = P(x=4) = C_4^4 p^4 (1-p)^0$$

Hence, once an estimate for p is obtained, the expected cell frequencies can be calculated using the above probabilities. Note that each of the 100 experiments consists of four trials and hence the complete experiment involves a total of 400 trials.

The best estimator of p is $\hat{p} = x/n$ (as in Chapter 9). Then,

$$\hat{p} = \frac{x}{n} = \frac{\text{number of successes}}{\text{number of trials}} = \frac{0(11) + 1(17) + 2(42) + 3(12) + 4(9)}{400} = \frac{1}{2}$$

The experiment (consisting of four trials) was repeated 100 times. There are a total of 400 trials in which the result "no successes in four trials" was observed 11 times, the result "one success in four trials" was observed 17 times, and so on. Then

$$p_0 = C_0^4 (1/2)^0 (1/2)^4 = 1/16 \qquad p_1 = C_1^4 (1/2)^1 (1/2)^3 = 4/16$$

$$p_2 = C_2^4 (1/2)^2 (1/2)^2 = 6/16 \qquad p_3 = C_3^4 (1/2)^3 (1/2)^1 = 4/16$$

$$p_4 = C_4^4 (1/2)^4 (1/2)^0 = 1/16$$

The observed and expected cell frequencies are shown in the following table.

x	0	1	2	3	4
O_i	11	17	42	21	9
E_i	6.25	25.00	37.50	25.00	6.25

and the statistic is

$$X^2 = \frac{(11 - 6.25)^2}{6.25} + \frac{(17 - 25.00)^2}{25.00} + \cdots + \frac{(9 - 6.25)^2}{6.25} = 8.56$$

In order to bound the p-value or set up a rejection region, it is necessary to determine the appropriate degrees of freedom associated with the test statistic. Two degrees of freedom are lost because:

1 The cell probabilities are restricted by the fact that $\sum p_i = 1$.

2 The binomial parameter p is unknown and must be estimated before calculating the expected cell counts. The number of degrees of freedom is equal to $k - 1 - 1 = k - 2 = 3$. With $df = 3$, the p-value for $X^2 = 8.56$ is between .025 and .05 and the null hypothesis can be rejected at the 5% level of significance. We conclude that the experiment in question does not fulfill the requirements for a binomial experiment.

14.51   The null hypothesis to be tested is

$$H_0 : p_1 = p_2 = p_3 = \frac{1}{3}$$

and the test statistic is

$$X^2 = \sum \frac{(O_i - E_i)^2}{E_i}$$

with $E_i = np_i = 200(1/3) = 66.67$ for $i = 1, 2, 3$. A table of observed and expected cell counts follows:

Entrance	1	2	3
O_i	83	61	56
E_i	66.67	66.67	66.67

Then

$$X^2 = \frac{(84-66.67)^2}{66.67} + \frac{(61-66.67)^2}{66.67} + \frac{(56-66.67)^2}{66.67} = 6.190$$

With $df = k-1 = 2$, the p-value is between .025 and .05 and we can reject H_0 at the 5% level of significance. There is a difference in preference for the three doors. A 95% confidence interval for p_1 is given as

$$\frac{x_1}{n} \pm z_{.025}\sqrt{\frac{\hat{p}_1\hat{q}_1}{n}} \Rightarrow \frac{83}{200} \pm 1.96\sqrt{\frac{.415(.585)}{200}} \Rightarrow .415 \pm .068$$

or $.347 < p_1 < .483$.

14.55 Since the cards for each of the three holidays will be either "humorous" or "not humorous", the table actually consists of two rows and three columns, and is shown with estimated expected and observed cell counts in the **Minitab** printout below.

Chi-Square Test: Fathers, Mothers, Valentines
```
Expected counts are printed below observed counts
Chi-Square contributions are printed below expected counts
        Fathers  Mothers  Valentines  Total
    1       100      125         120     345
         115.00   115.00      115.00
          1.957    0.870       0.217
    2       400      375         380    1155
         385.00   385.00      385.00
          0.584    0.260       0.065
Total       500      500         500    1500
Chi-Sq = 3.953, DF = 2, P-Value = 0.139
```
The test statistic for the equality of the three population proportions is

$X^2 = 3.953$ with p-value $= .139$ and H_0 is not rejected. There is insufficient evidence to indicate a difference in the proportion of humorous cards for the three holidays.

14.59 **a** The 2×3 contingency table is analyzed as in previous exercises. The **Minitab** printout below shows the observed and estimated expected cell counts, the test statistic and its associated p-value.

Chi-Square Test: 3 or fewer, 4 or 5, 6 or more
```
Expected counts are printed below observed counts
Chi-Square contributions are printed below expected counts
          3 or
          fewer  4 or 5  6 or more  Total
    1        49      43         34    126
          37.89   42.63      45.47
           3.254   0.003      2.895
    2        31      47         62    140
          42.11   47.37      50.53
           2.929   0.003      2.605
Total        80      90         96    266
Chi-Sq = 11.690, DF = 2, P-Value = 0.003
```

The results are highly significant (p-value $= .003$) and we conclude that there is a difference in the susceptibility to colds depending on the number of relationships you have.

b The proportion of people with colds is calculated conditionally for each of the three groups, and is shown in the table below.

	Three or fewer	Four or five	Six or more
Cold	$\dfrac{49}{80} = .61$	$\dfrac{43}{90} = .48$	$\dfrac{34}{96} = .35$
No cold	$\dfrac{31}{80} = .39$	$\dfrac{47}{90} = .52$	$\dfrac{62}{96} = .65$
Total	1.00	100	1.00

As the researcher suspects, the susceptibility to a cold seems to decrease as the number of relationships increases!

14.65 Use the first **Goodness-of-Fit** applet. Enter the observed values into the three cells in the first row, and the expected cell counts will automatically appear in the second row. The value of $X^2 = 18.5$ with p-value $= .0001$ provide sufficient evidence to reject H_0 and conclude that customers have a preference for one of the three brands (in this case, Brand II).

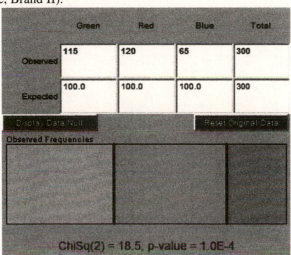

14.69 The data is analyzed as a 2×3 contingency table with estimated expected cell counts shown in parentheses. Use the **Chi-Square Test of Independence** applet. Your results should agree with the hand calculations shown below.

Party	Opinion			Total
	1	2	3	
Republican	114	53	17	184
	(120.86)	(48.10)	(15.03)	
Democrat	87	27	8	122
	(80.14)	(31.89)	(9.97)	
Total	201	80	25	306

The test statistic is

$$X^2 = \frac{(114-120.86)^2}{120.86} + \frac{(53-48.10)^2}{48.10} + \cdots + \frac{(8-9.97)^2}{9.97} = 2.87$$

With $df = 2$, the p-value is greater than .10 (the applet reports p-value $= .2378$) and H_0 is not rejected. There is no evidence that party affiliation has any effect on opinion.

15: Nonparametric Statistics

15.1 **a** If distribution 1 is shifted to the right of distribution 2, the rank sum for sample 1 (T_1) will tend to be large. The test statistic will be T_1^*, the rank sum for sample 1 if the observations had been ranked from large to small. The null hypothesis will be rejected if T_1^* is unusually small.

b From Table 7a with $n_1 = 6$, $n_2 = 8$ and $\alpha = .05$, H_0 will be rejected if $T_1^* \le 31$.

c From Table 7c with $n_1 = 6$, $n_2 = 8$ and $\alpha = .01$, H_0 will be rejected if $T_1^* \le 27$.

15.5 If H_a is true and population 1 lies to the right of population 2, then T_1 will be large and T_1^* will be small. Hence, the test statistic will be T_1^* and the large sample approximation can be used. Calculate

$$T_1^* = n_1(n_1 + n_2 + 1) - T_1 = 12(27) - 193 = 131$$

$$\mu_T = \frac{n_1(n_1 + n_2 + 1)}{2} = \frac{12(26 + 1)}{2} = 162$$

$$\sigma_T^2 = \frac{n_1 n_2 (n_1 + n_2 + 1)}{12} = \frac{12(14)(27)}{12} = 378$$

The test statistic is

$$z = \frac{T_1 - \mu_T}{\sigma_T} = \frac{131 - 162}{\sqrt{378}} = -1.59$$

The rejection region with $\alpha = .05$ is $z < -1.645$ and H_0 is not rejected. There is insufficient evidence to indicate a difference in the two population distributions.

15.9 Similar to previous exercises. The data, with corresponding ranks, are shown in the following table.

Deaf (1)	Hearing (2)
2.75 (15)	0.89 (1)
2.14 (11)	1.43 (7)
3.23 (18)	1.06 (4)
2.07 (10)	1.01 (3)
2.49 (14)	0.94 (2)
2.18 (12)	1.79 (8)
3.16 (17)	1.12 (5.5)
2.93 (16)	2.01 (9)
2.20 (13)	1.12 (5.5)
$T_1 = 126$	

Calculate

$$T_1 = 126$$

$$T_1^* = n_1 (n_1 + n_2 + 1) - T_1 = 9(19) - 126 = 45$$

The test statistic is

$$T = \min\left(T_1, T_1^*\right) = 45$$

With $n_1 = n_2 = 9$, the two-tailed rejection region with $\alpha = .05$ is found in Table 7b to be $T_1^* \leq 62$. The observed value, $T = 45$, falls in the rejection region and H_0 is rejected. We conclude that the deaf children do differ from the hearing children in eye-movement rate.

15.13 **a** If a paired difference experiment has been used and the sign test is one-tailed $\left(H_a : p > .5\right)$, then the experimenter would like to show that one population of measurements lies above the other population. An exact practical statement of the alternative hypothesis would depend on the experimental situation.

b It is necessary that α (the probability of rejecting the null hypothesis when it is true) take values less than $\alpha = .15$. Assuming the null hypothesis to be true, the two populations are identical and consequently,

$p = P\left(A \text{ exceeds B for a given pair of observations}\right)$ is 1/2. The binomial probability was discussed in Chapter 5. In particular, it was noted that the distribution of the random variable x is symmetrical about the mean np when $p = 1/2$. For example, with $n = 25$, $P(x = 0) = P(x = 25)$. Similarly, $P(x = 1) = P(x = 24)$ and so on. Hence, the lower tailed probabilities tabulated in Table 1, Appendix I will be identical to their upper tailed equivalent probabilities. The values of α available for this upper tailed test and the corresponding rejection regions are shown below.

Rejection Region	α
$x \geq 20$	.002
$x \geq 19$	.007
$x \geq 18$	.022
$x \geq 17$	.054
$x \geq 16$	.115

15.17 **a** If assessors A and B are equal in their property assessments, then p, the probability that A's assessment exceeds B's assessment for a given property, should equal 1/2. If one of the assessors tends to be more conservative than the other, then either $p > 1/2$ or $p < 1/2$. Hence, we can test the equivalence of the two assessors by testing the hypothesis

$$H_0 : p = 1/2 \quad \text{versus} \quad H_a : p \neq 1/2$$

using the test statistic x, the number of times that assessor A exceeds assessor B for a particular property assessment. To find a two-tailed rejection region with α close to .05, use Table 1 with $n = 8$ and $p = .5$. For the rejection region $\{x = 0, x = 8\}$ the

value of α is $.004 + .004 = .008$, while for the rejection region $\{x = 0,1,7,8\}$ the value of α is $.035 + .035 = .070$ which is closer to .05. Hence, using the rejection region $\{x \leq 1 \text{ or } x \geq 7\}$, the null hypothesis is not rejected, since $x =$ number of properties for which A exceeds B = 6. The p-value for this two-tailed test is
$$p\text{-value} = 2P(x \geq 6) = 2(1 - .855) = .290$$

Since the p-value is greater than .10, the results are not significant; H_0 is not rejected (as with the critical value approach).

b The t statistic used in Exercise 10.45 allows the experimenter to reject H_0, while the sign test fails to reject H_0. This is because the sign test used less information and makes fewer assumptions than does the t test. If all normality assumptions are met, the t test is the more powerful test and can reject when the sign test cannot.

15.21 **a** H_0: population distributions 1 and 2 are identical
H_a: the distributions differ in location

b Since Table 8, Appendix I gives critical values for rejection in the lower tail of the distribution, we use the smaller of T^+ and T^- as the test statistic.

c From Table 8 with $n = 30$, $\alpha = .05$ and a two-tailed test, the rejection region is $T \leq 137$.

d Since $T^+ = 249$, we can calculate
$$T^- = \frac{n(n+1)}{2} - T^+ = \frac{30(31)}{2} - 249 = 216.$$

The test statistic is the smaller of T^+ and T^- or $T = 216$ and H_0 is not rejected. There is no evidence of a difference between the two distributions.

15.25 **a** The hypothesis to be tested is

H_0: population distributions 1 and 2 are identical
H_a: the distributions differ in location

and the test statistic is T, the rank sum of the positive (or negative) differences. The ranks are obtained by ordering the differences according to their absolute value. Define d_i to be the difference between a pair in populations 1 and 2 (i.e., $x_{1i} - x_{2i}$). The differences, along with their ranks (according to absolute magnitude), are shown in the following table.

d_i	.1	.7	.3	−.1	.5	.2	.5
Rank $|d_i|$	1.5	7	4	1.5	5.5	3	5.5

The rank sum for positive differences is $T^+ = 26.5$ and the rank sum for negative differences is $T^- = 1.5$ with $n = 7$. Consider the smaller rank sum and determine the appropriate lower portion of the two-tailed rejection region. Indexing $n = 7$ and $\alpha = .05$ in Table 8, the rejection region is $T \leq 2$ and H_0 is rejected. There is a difference in the two population locations.

b The results do not agree with those obtained in Exercise 15.16. We are able to reject H_0 with the more powerful Wilcoxon test.

15.29 **a** The paired data are given in the exercise. The differences, along with their ranks (according to absolute magnitude), are shown in the following table.

d_i	1	2	–1	1	3	1	–1	3	–2	3	1	0
Rank $|d_i|$	3.5	7.5	3.5	3.5	10	3.5	3.5	10	7.5	10	2.5	-

Let $p = P(\text{A exceeds B for a given intersection})$ and $x =$ number of intersections at which A exceeds B. The hypothesis to be tested is

$$H_0 : p = 1/2 \quad \text{versus} \quad H_a : p \neq 1/2$$

using the sign test with x as the test statistic.

Critical value approach: Various two tailed rejection regions are tried in order to find a region with $\alpha \approx .05$. These are shown in the following table.

Rejection Region	α
$x \leq 1; x \geq 10$	.012
$x \leq 2; x \geq 9$	.066
$x \leq 3; x \geq 8$	.226

We choose to reject H_0 if $x \leq 2$ or $x \geq 9$ with $\alpha = .066$. Since $x = 8$, H_0 is not rejected. There is insufficient evidence to indicate a difference between the two methods.

p-value approach: For the observed value $x = 8$, calculate the two-tailed p-value:

$$p\text{-value} = 2P(x \geq 8) = 2(1 - .887) = .226$$

Since the p-value is greater than .10, H_0 is not rejected.

b To use the Wilcoxon signed rank test, we use the ranks of the absolute differences shown in the table above. Then $T^+ = 51.5$ and $T^- = 14.5$ with $n = 11$. Indexing $n = 11$ and $\alpha = .05$ in Table 8, the lower portion of the two-tailed rejection region is $T \leq 11$ and H_0 is not rejected, as in part **a**.

15.31 **a** Since the experiment has been designed as a paired experiment, there are three tests available for testing the differences in the distributions with and without imagery – (1) the paired difference t test; (2) the sign test or (3) the Wilcoxon signed rank test. In order to use the paired difference t test, the scores must be approximately normal; since the number of words recalled has a binomial distribution with $n = 25$ and unknown recall probability, this distribution may not be approximately normal.

b Using the **sign test**, the hypothesis to be tested is

$$H_0 : p = 1/2 \quad \text{versus} \quad H_a : p > 1/2$$

For the observed value $x = 0$ we calculate the two-tailed p-value:

$$p\text{-value} = 2P(x \leq 0) = 2(.000) = .000$$

The results are highly significant; H_0 is rejected and we conclude there is a difference in the recall scores with and without imagery.

Using the **Wilcoxon signed-rank test**, the differences will all be positive ($x = 0$ for the sign test), so that and

$$T^+ = \frac{n(n+1)}{2} = \frac{20(21)}{2} = 210 \quad \text{and} \quad T^- = 210 - 210 = 0$$

Indexing $n = 20$ and $\alpha = .01$ in Table 8, the lower portion of the two-tailed rejection region is $T \le 37$ and H_0 is rejected.

15.35   The data with corresponding ranks in parentheses are shown below.

	Age		
10 – 19	20 – 39	40 – 59	60 – 69
29 (21)	24 (8)	37 (39)	28 (18)
33 (29.5)	27 (15)	25 (10.5)	29 (21)
26 (12.5)	33 (29.5)	22 (5.5)	34 (34)
27 (15)	31 (24)	33 (29.5)	36 (37.5)
39 (40)	21 (3)	28 (18)	21 (3)
35 (36)	28 (18)	26 (12.5)	20 (1)
33 (29.5)	24 (8)	30 (23)	25 (10.5)
29 (21)	34 (34)	34 (34)	24 (8)
36 (37.5)	21 (3)	27 (15)	33 (29.5)
22 (5.5)	32 (25.5)	33 (29.5)	32 (25.5)
$T_1 = 247.5$	$T_2 = 168$	$T_3 = 216.5$	$T_4 = 188$
$n_1 = 10$	$n_2 = 10$	$n_3 = 10$	$n_4 = 10$

a   The test statistic, based on the rank sums, is

$$H = \frac{12}{n(n+1)} \sum \frac{T_i^2}{n_i} - 3(n+1)$$

$$= \frac{12}{40(41)} \left[\frac{(247.5)^2}{10} + \frac{(168)^2}{10} + \frac{(216.5)^2}{10} + \frac{(188)^2}{10} \right] - 3(41) = 2.63$$

The rejection region with $\alpha = .01$ and $k - 1 = 3$ df is based on the chi-square distribution, or $H > \chi_{.01}^2 = 11.35$. The null hypothesis is not rejected. There is no evidence of a difference in location.

b   Since the observed value $H = 2.63$ is less than $\chi_{.10}^2 = 6.25$, the p-value is greater than .10.

c-d   From Exercise 11.60, $F = .87$ with 3 and 36 df. Again, the p-value is greater than .10 and the results are the same.

15.39   The ranks of the data are shown on the next page.

	Treatment			
Block	1	2	3	4
1	4	1	2	3
2	4	1.5	1.5	3
3	4	1	3	2
4	4	1	2	3
5	4	1	2.5	2.5
6	4	1	2	3
7	4	1	3	2
8	4	1	2	3
	$T_1 = 32$	$T_2 = 8.5$	$T_3 = 18$	$T_4 = 21.5$

a The test statistic is

$$F_r = \frac{12}{bk(k+1)} \sum T_i^2 - 3b(k+1)$$

$$= \frac{12}{8(4)(5)} \left[(32)^2 + (8.5)^2 + 18^2 + (21.5)^2 \right] - 3(8)(5) = 21.19$$

and the rejection region is $F_r > \chi^2_{.05} = 7.81$. Hence, H_0 is rejected and we conclude that there is a difference among the four treatments.

b The observed value, $F_r = 21.19$, exceeds $\chi^2_{.005}$, p-value $< .005$.

c-e The analysis of variance is performed as in Chapter 11. The ANOVA table is shown below.

Source	df	SS	MS	F
Treatments	3	198.34375	66.114583	75.43
Blocks	7	220.46875	31.495536	
Error	21	18.40625	0.876488	
Total	31	437.40625		

The analysis of variance F test for treatments is $F = 75.43$ and the approximate p-value with 3 and 21 df is p-value $< .005$. The result is identical to the parametric result.

15.43 Table 9, Appendix I gives critical values r_0 such that $P(r_s \geq r_0) = \alpha$. Hence, for an upper-tailed test, the critical value for rejection can be read directly from the table.
a $r_s \geq .425$ **b** $r_s \geq .601$

15.47 **a** The two variables (rating and distance) are ranked from low to high, and the results are shown in the following table.

Voter	x	y	Voter	x	y
1	7.5	3	7	6	4
2	4	7	8	11	2
3	3	12	9	1	10
4	12	1	10	5	9
5	10	8	11	9	5.5
6	7.5	11	12	2	5.5

Calculate $\quad \sum x_i y_i = 442.5 \quad \sum x_i^2 = 649.5 \quad \sum y_i^2 = 649.5$

$\quad\quad\quad\quad\quad\quad n = 12 \quad\quad\quad \sum x_i = 78 \quad\quad \sum y_i = 78$

Then

$$S_{xy} = 422.5 - \frac{78^2}{12} = -84.5 \quad\quad S_{xx} = 649.5 - \frac{78^2}{12} = 142.5$$

$$S_{yy} = 649.5 - \frac{78^2}{12} = 142.5$$

and

$$r_s = \frac{S_{xy}}{\sqrt{S_{xx}S_{yy}}} = \frac{-84.5}{142.5} = -.593 \ .$$

b The hypothesis of interest is H_0: no correlation versus H_a: negative correlation. Consulting Table 9 for $\alpha = .05$, the critical value of r_s, denoted by r_0 is $-.497$. Since the value of the test statistic is less than the critical value, the null hypothesis is rejected. There is evidence of a significant negative correlation between rating and distance.

15.51 Refer to Exercise 15.50. To test for positive correlation with $\alpha = .05$, index .05 in Table 9 and the rejection region is $r_s \geq .600$. We reject the null hypothesis of no association and conclude that a positive correlation exists between the teacher's ranks and the ranks of the IQs.

15.55 **a** Define $p = P(\text{response for stimulus 1 exceeds that for stimulus 2})$ and $x = $ number of times the response for stimulus 1 exceeds that for stimulus 2. The hypothesis to be tested is

$$H_0 : p = 1/2 \quad \text{versus} \quad H_a : p \neq 1/2$$

using the sign test with x as the test statistic. Notice that for this exercise $n = 9$, and the observed value of the test statistic is $x = 2$. Various two tailed rejection regions are tried in order to find a region with $\alpha \approx .05$. These are shown in the following table.

Rejection Region	α
$x = 0; x = 9$	.004
$x \leq 1; x \geq 8$	.040
$x \leq 2; x \geq 7$	.180

We choose to reject H_0 if $x \leq 1$ or $x \geq 8$ with $\alpha = .040$. Since $x = 2$, H_0 is not rejected. There is insufficient evidence to indicate a difference between the two stimuli.

b The experiment has been designed in a paired manner, and the paired difference test is used. The differences are shown below.

$$d_i \quad -.9 \quad -1.1 \quad 1.5 \quad -2.6 \quad -1.8 \quad -2.9 \quad -2.5 \quad 2.5 \quad -1.4$$

The hypothesis to be tested is

$$H_0 : \mu_1 - \mu_2 = 0 \quad\quad H_a : \mu_1 - \mu_2 \neq 0$$

Calculate

$$\bar{d} = \frac{\sum d_i}{n} = \frac{-9.2}{9} = -1.022$$

$$s_d^2 = \frac{\sum d_i^2 - \frac{(\sum d_i)^2}{n}}{n-1} = \frac{37.14 - 9.404}{8} = 3.467$$

and the test statistic is

$$t = \frac{\bar{d}}{\sqrt{\frac{s_d^2}{n}}} = \frac{-1.022}{\sqrt{\frac{3.467}{9}}} = -1.646$$

The rejection region with $\alpha = .05$ and 8 df is $|t| > 2.306$ and H_0 is not rejected.

15.59   The data, with corresponding ranks, are shown in the following table.

A (1)	B (2)
6.1 (1)	9.1 (16)
9.2 (17)	8.2 (8)
8.7 (12)	8.6 (11)
8.9 (13.5)	6.9 (2)
7.6 (5)	7.5 (4)
7.1 (3)	7.9 (7)
9.5 (18)	8.3 (9.5)
8.3 (9.5)	7.8 (6)
9.0 (1.5)	8.9(13.5)
$T_1 = 94$	

The difference in the brightness levels using the two processes can be tested using the nonparametric Wilcoxon rank sum test, or the parametric two-sample t test.

1 To test the null hypothesis that the two population distributions are identical, calculate

$$T_1 = 1 + 17 + \cdots + 1.5 = 94$$
$$T_1^* = n_1(n_1 + n_2 + 1) - T_1 = 9(18+1) - 94 = 77$$

The test statistic is

$$T = \min(T_1, T_1^*) = 77$$

With $n_1 = n_2 = 9$, the two-tailed rejection region with $\alpha = .05$ is found in Table 7b to be $T_1^* \leq 62$. The observed value, $T = 77$, does not fall in the rejection region and H_0 is not rejected. We cannot conclude that the distributions of brightness measurements is different for the two processes.

2 To test the null hypothesis that the two population means are identical, calculate

$$\bar{x}_1 = \frac{\sum x_{1j}}{n_1} = \frac{74.4}{9} = 8.2667 \qquad \bar{x}_2 = \frac{\sum x_{2j}}{n_2} = \frac{73.2}{9} = 8.1333$$

$$s^2 = \frac{(n_1-1)s_1^2 + (n_2-1)s_2^2}{n_1+n_2-2} = \frac{625.06 - \frac{(74.4)^2}{9} + 599.22 - \frac{(73.2)^2}{9}}{16} = .8675$$

and the test statistic is

$$t = \frac{\bar{x}_1 - \bar{x}_2}{\sqrt{s^2\left(\frac{1}{n_1}+\frac{1}{n_2}\right)}} = \frac{8.27 - 8.13}{\sqrt{.8675\left(\frac{2}{9}\right)}} = .304$$

The rejection region with $\alpha = .05$ and 16 degrees of freedom is $|t| > 1.746$ and H_0 is not rejected. There is insufficient evidence to indicate a difference in the average brightness measurements for the two processes. Notice that the nonparametric and parametric tests reach the same conclusions.

15.61 Since this is a paired experiment, you can choose either the sign test, the Wilcoxon signed rank test, or the parametric paired t test. Since the tenderizers have been scored on a scale of 1 to 10, the parametric test is not applicable. Start by using the easiest of the two nonparametric tests – the sign test.

Define $p = P(\text{tenderizer A exceeds B for a given cut})$ and $x =$ number of times that A exceeds B. The hypothesis to be tested is

$$H_0 : p = 1/2 \quad \text{versus} \quad H_a : p \neq 1/2$$

using the sign test with x as the test statistic. Notice that for this exercise $n = 8$ (there are two ties), and the observed value of the test statistic is $x = 2$.

p-value approach: For the observed value $x = 2$, calculate

$$p\text{-value} = 2P(x \le 2) = 2(.145) = .290$$

Since the p-value is greater than .10, H_0 is not rejected. There is insufficient evidence to indicate a difference between the two tenderizers.

If you use the Wilcoxon signed rank test, you will find $T^+ = 7$ and $T^- = 29$ which will not allow rejection of H_0 at the 5% level of significance. The results are the same.

15.65 The hypothesis to be tested is

H_0: population distributions 1 and 2 are identical
H_a: the distributions differ in location

and the test statistic is T, the rank sum of the positive (or negative) differences. The ranks are obtained by ordering the differences according to their absolute value. Define d_i to be the difference between a pair in populations 1 and 2 (i.e., $x_{1i} - x_{2i}$).

The differences, along with their ranks (according to absolute magnitude), are shown in the following table.

d_i	−31	−31	−6	−11	−9	−7	7
Rank $|d_i|$	14.5	14.5	4.5	12.5	10.5	7	7

d_i	−11	7	−9	−2	−8	−1	−6	−3
Rank $|d_i|$	12.5	7	10.5	2	9	1	4.5	3

The rank sum for positive differences is $T^+ = 14$ and the rank sum for negative differences is $T^- = 106$ with $n = 15$. Consider the smaller rank sum and determine the appropriate lower portion of the two-tailed rejection region. Indexing $n = 15$ and $\alpha = .05$ in Table 8, the rejection region is $T \le 25$ and H_0 is rejected. We conclude that there is a difference between math and art scores.

15.69 **a-b** Since the experiment is a completely randomized design, the Kruskal Wallis H test is used. The combined ranks are shown below.

Plant	Ranks					T_i
A	9	12	5	1	7	34
B	11	15	4	19	14	63
C	3	13	2	9	6	33
D	20	17	9	16	18	80

The test statistic, based on the rank sums, is

$$H = \frac{12}{n(n+1)} \sum \frac{T_i^2}{n_i} - 3(n+1)$$

$$= \frac{12}{20(21)} \left[\frac{(34)^2}{5} + \frac{(63)^2}{5} + \frac{(33)^2}{5} + \frac{(80)^2}{5} \right] - 3(21) = 9.08$$

With $df = k - 1 = 3$, the observed value $H = 9.08$ is between $\chi_{.025}$ and $\chi_{.05}$ so that $.025 < p\text{-value} < .05$. The null hypothesis is rejected and we conclude that there is a difference among the four plants.
c From Exercise 11.66, $F = 5.20$, and H_0 is rejected. The results are the same.

15.73 The data are already in rank form. The "substantial experience" sample is designated as sample 1, and $n_1 = 5, n_2 = 7$. Calculate

$$T_1 = 19$$

$$T_1^* = n_1(n_1 + n_2 + 1) - T_1 = 5(13) - 19 = 46$$

The test statistic is

$$T = \min(T_1, T_1^*) = 19$$

With $n_1 = n_2 = 12$, the one-tailed rejection region with $\alpha = .05$ is found in Table 7a to be $T_1 \le 21$. The observed value, $T = 19$, falls in the rejection region and H_0 is rejected. There is sufficient evidence to indicate that the review board considers experience a prime factor in the selection of the best candidates.

15.77 The data with corresponding ranks in parentheses are shown on the next page.

Training Periods (hours)			
.5	1.0	1.5	2.0
8 (9.5)	9 (11.5)	4 (1.5)	4 (1.5)
14 (14)	7 (7)	6 (5)	7 (7)
9 (11.5)	5 (3.5)	7 (7)	5 (3.5)
12 (13)		8 (9.5)	
$T_1 = 48$	$T_2 = 22$	$T_3 = 23$	$T_4 = 12$
$n_1 = 4$	$n_2 = 3$	$n_3 = 4$	$n_4 = 3$

The test statistic, based on the rank sums, is

$$H = \frac{12}{n(n+1)} \sum \frac{T_i^2}{n_i} - 3(n+1)$$

$$= \frac{12}{14(15)} \left[\frac{(48)^2}{4} + \frac{(22)^2}{3} + \frac{(23)^2}{4} + \frac{(12)^2}{3} \right] - 3(15) = 7.4333$$

The rejection region with $\alpha = .01$ and $k - 1 = 3$ df is based on the chi-square distribution, or $H > \chi^2_{.01} = 11.34$. The null hypothesis is not rejected and we conclude that there is insufficient evidence to indicate a difference in the distribution of times for the four groups.